Floating Drilling: Equipment and Its Use

Gulf Publishing Company
Book Division
Houston, London, Paris, Tokyo

Riley Sheffield

Floating Drilling:
Equipment and Its Use

Practical Drilling Technology · Volume 2

Floating Drilling:
Equipment and Its Use

Library of Congress Cataloging in Publication Data
Sheffield, Riley, 1931–
    Floating drilling, equipment and its use.
    (Practical drilling technology; v.2)
    Includes bibliographical references and index.
    1. Oil well drilling, Submarine.
    2. Drilling platforms.
    I. Title.
TN871.3.S53   627'.98   80-14311
ISBN 0-87201-289-1

First Printing, June 1980
Second Printing, November 1982

Contents

Hydraulic Control 112, Components 112, Regulations 115, Control Pods 116, Accumulators 116, Shuttle Valves 116, Lines on the Stack 116, Hydraulic Unit 117, Redundant Operation 118, Operating Pressures and Capacities 122, Accumulator Capacity Requirements 125, Pumping Capacity 127, Factors Influencing Reaction Time 128, BOP Testing 130, Surface Testing 130, Subsea Testing 131, Pressure Integrity of the BOP Stack 131, Backup for BOP Control 133, References 136

Riser Components 139, Riser Joints 139, Ball Joints 140, Slip Joints 140, Diverters 142, Jumper Lines 143, Buoyancy Modules 144, Riser Tensioning 144, Non-Operational but Connected Mode 148, Operational Mode 149, Riser Appraisal 150, Steel Properties and Fabrication Practices 150, Fatigue 151, Operating Factors 151, Kill/Choke Lines 153, Inspection and Maintenance 157, Instrumentation to Protect the Riser 158, An Operating Procedure 159, Guidance Systems 160, Guidelines 160, Reentry 162, References 164

Passive Motion Compensation 167, Tensioners 172, Heave Compensators 175, Traveling Block Compensators 176, Crown Block Compensators 179, Active and Semiactive Compensation 180, Bumper Subs 182, References 184

Downhole Test Equipment 186, Test Tools 186, Slip Joint Safety Valve 187, Volume-Pressure Balanced Slip Joints 188, Reverse Circulating Subs 190, Sea Floor Shut In 192, Space Out 193, Surface Test Equipment 198, References 202

Location Determination 203, Radio Positioning Systems 203, Navy Navigational Satellite System 208, Workboats 210, Loading 210, Arrangement 213, Anchor Handling and Towing 214, Other Workboat Requirements 215, Stability 215, Inspection 217, Tides 218, References 222

Acknowledgments

"WOW! And I don't mean Wait'n-On-Weather."* Individual acknowledgments would take two volumes the size of this one. I have yet to meet a roughneck, salesman, engineer, driller, manager, drilling superintendent, technician, lawyer, geologist—or anyone involved in the industry—who did not deepen my insight into drilling operations and help pave the way for this book. Their personalities and creativity have allowed the oil industry to operate where such operations were once thought impossible. And I must acknowledge one of the greatest teachers of all time to those who are willing to learn:

THE OIL PATCH

*This sentence published under the protest of my editor.

Preface

In this book we will look closely at the components of the operation of drilling a well from a vessel floating on the water. These components include the vessel itself, the drilling equipment, the mooring system, wellheads, blowout preventors and others. I speak of components, but it is important to remember that the rig must operate effectively as a *unit*. And so, its components must not only meet individual specifications; they must also be capable of operating with other components of the drilling rig.

When choosing a rig, inspections should be made using a check list of important items. This list and its contents will vary from company to company. The check list should contain enough information to evaluate a vessel and to be of value for future use in the drilling office. Appendix A contains an example of such a check list for evaluation of a floating drilling vessel.

The purpose of a drilling rig is the same both on land and at sea: to drill and to complete a well successfully and economically. The major differences between a floating drilling vessel and a land rig are the vessel itself, the equipment to compensate for the vessel motion, and a wellhead that is located on the seabed. These differences require a somewhat drastic change in attitude, because the actual drilling equipment alone is not necessarily the most expensive item. In many instances it may be cheaper to change out some of the drilling equipment rather than to jack it up and float a new vessel under it.

To a land drilling supervisor, the discussion immediately following may seem extraneous, as several subjects are covered before actual drilling equipment is discussed. This is because a basic knowledge of the vessel itself is prerequisite to a detailed discussion of the drilling equipment to be used. The choice and use of a drilling vessel is contingent on its motion characteristics, stability, and station keeping ability, as well as on the equipment needed to "make hole."

As opposed to a discussion of hydraulics, mud properties and the choice of bits, the primary objective of this book is to discuss the difference between land drilling and floating drilling, and the choice of equipment for floating drilling.

To my beloved wife, Lou,
and to our children—
especially Louise,
who helped with the manuscript

Floating Drilling: Equipment and Its Use

Drilling a Well from a Floating Vessel

The following discussion is based on a typical, or perhaps an ideal, exploration well to be drilled from a floating drilling vessel. Casing sizes and depths vary between operators and with areas of operation: however, the sequence of operations discussed is that of most exploration wells drilled from a floater. Those who are familiar with land drilling or with floating drilling will realize that many drilling problems are only mentioned or are omitted entirely.

Let us suppose that a rig has just completed a well and is underway to a new well site. This is a very busy time on a drilling vessel. The rig crew is getting equipment ready for the next well. The hole opener for drilling the 36-inch hole to house the structural 30-inch casing must be pulled from storage and prepared for use. The 30-inch casing must be measured and brought on deck for running. Some equipment specialists are on board to check the riser and blowout preventers (BOP) systems. The riser joints are being visually inspected and prepared for running. Some of the joints will probably be sent ashore for a routine but very thorough inspection. The water depth at the new location is known and the tide tables are available so that the correct number of riser joints can be prepared for running. The blowout preventers (BOPs) are at the surface where they can be main-tained. Usually, the BOPs are on the sea floor, but now they are being attacked by specialists dedicated to the maintenance of special parts of the stack. The BOP stack is essential to well control and must function prop-erly during the drilling operation. If a kick had to be circulated out during the previous well, the choke manifold will probably be disassembled and critical areas will be checked for erosion, just as on a land rig.

While the rig crew is occupied with the drilling equipment, the ship's crew is operating the vessel and preparing to run anchors. At the location, the site has been surveyed and marked with buoys, and workboats (also called "anchorboats") are standing by to assist in setting anchors. The vessel moves into position, the heading is adjusted, and running the anchors begins. When sufficient anchors have been run, the anchors are pretensioned, the lines are slacked off to a predetermined operating tension and the final adjustment of the vessel position and heading are made. Anchoring up requires from six hours to over three days.

Now the well can be spudded. First, a temporary guide base is lowered to the seabed. This massive piece of steel will be used as a template to guide the drilling equipment and the first string of casing into the well via attached guidelines. The 36-inch hole is drilled 80 to 200 feet below the mudline and the 30-inch casing is run and cemented in place. This casing is strictly for structural support and will not sustain any pressure. A permanent guide frame to support the wellhead and BOP stack is run on the 30-inch casing.

Next, a 26-inch hole is drilled for a 20-inch conductor pipe that will be set about 1,000 feet below the mudline. Various operators require running the riser (without the BOP stack) and a diverter so that if shallow gas sands are encountered, the well can be brought under control with minimum danger to the personnel and equipment. The diverter keeps the mud and well effluent from hindering work on the rig floor. It is needed because the well must not be shut in on the 30-inch casing. Before the riser is run, the casing shoe is drilled with a 26-inch hole opener. After the riser is run, the hole will be drilled and underreamed to 26 inches. If the riser is not run, the entire hole will be drilled with a 26-inch bit.

Before running the casing, the mud weight is increased enough to allow for the decrease in hydrostatic pressure that accompanies pulling the riser. Then the riser is pulled. The 20-inch casing is run with the wellhead, and is cemented in place. While waiting for the cement to set, the stack and riser will be rigged (nippled up), run, and latched onto the wellhead. The BOPs should never be closed on the conductor pipe. The reason the BOPs are run now, as opposed to running the riser only, is to avoid having to make a "round trip" with the riser later. If the weather and the operation have gone smoothly, about five days will have lapsed since first arriving on location.

Next, a 17½-inch hole is drilled for the 13⅜-inch surface casing. The surface casing seals off the shallow, low-pressure sands near the surface so that a kick can be circulated out. For the first time since spudding, good drilling practices can be approached, such as hydraulics, bit weight, rotary speed, mud properties, and annular velocity. This hole may be drilled with a nominal 17½-inch bit or it may be drilled and underreamed to 17½-inch depending on the size of the wellhead and BOPs used. Some operators prefer to drill a nominal 12¼-inch pilot hole, log, and then underream. This gives additional geologic information about the formations penetrated; it also slows down the drilling process.

After the hole has been drilled, the surface casing is run and hung in the wellhead. The casing is cemented, using subsea cementing plugs. These plugs are located at the top of the casing and are launched remotely from the surface. After the cement has set, a casing seal is run to seal off the annulus between the 13⅜-inch and the 20-inch casings. These seals must be tested before the casing shoe is drilled. The surface casing will be set 3,000 to 5,000 feet below the mudline, and the time required for drilling the hole will vary, depending on the drilling practices and the problems that may be encountered. Three to as many as six weeks may elapse following first arrival on location.

After setting the surface casing, many governments require a leak-off test. This is a test to estimate the fracture gradient of the formation just below the casing shoe. The operational value of this test is debatable and some operators prefer not to run them. Whether or not a leak-off test is run, the next hole drilled will be a nominal 12¼-inch for the 9⅝-inch casing. Ideally, the operator would like to set the 9⅝-inch just above a good producing zone. This is often wishful thinking, because the probability of finding any oil in an exploration well is less than 10%. Drilling problems, such as abnormally high or abnormally low pressure formations that may require setting the casing or supplemental liners at unanticipated depths, may also occur.

If the operator is lucky, the 9⅝-inch casing will be set right above a producing formation. The casing will be set, cemented, sealed and tested just as the surface casing string was.

Next, a nominal 8¼-inch hole will be drilled for a 7-inch production casing string or a production liner. After the casing or liner has been set, the well will be tested.

The major reason that the well is drilled is to determine if economically recoverable petroleum reserves are located at that site. Well testing is of prime importance to this evaluation. The cement jobs, the casing seals, the operation of the BOPs and the performance of the rig personnel also get tested with the well. Up to now, two months or more may have elapsed since arriving on location. Testing may take from two days to three weeks.

After testing, the well is often plugged and abandoned (P&A). The well is plugged by squeezing cement into the perforations used for testing and then spotting cement at given intervals in the casing as the drill string is pulled from the well for the last time. Now the riser and BOP stack can be retrieved. The wellhead equipment, including the temporary and permanent guide bases, is retrieved by cutting the casing about 10 feet below the wellhead. Either explosives or mechanical cutters may be used. Once the wellhead has been retrieved, the process of pulling the anchors begins. Soon, the vessel will begin the move to another location, and the preparation for the next well has already begun.

The number of wells that a floater will drill in one year varies depending on transit times, the drilling equipment, the crew, the weather and the problems encountered at the different well sites. Some floaters have drilled less than three wells in a year while others have drilled as many as 14. When drilling rank wildcats from a floater, especially in hostile areas, four to five wells per year seems to be a reasonable average.

2

The Vessel

There are two basic types of floating drilling vessels. These are *ship-like vessels* that look like a normal vessel or barge with a drilling rig on it; and the *semi-submersibles*, also called column stabilized units or "semis." The semis are unique and have been developed specifically for drilling in choppy seas. Semis show no similarity to a conventional ship (see Figure 2-1).

Barge rigs are usually used in protected areas and are generally very economical. The drill ships are self propelled and may be either moored or dynamically positioned over the well. Beginning in about 1970, essentially all semis were built with thrusters to assist in towing and positioning the platform for mooring. Semis built after 1973 usually have some type of self propulsion, and one semi with dynamic positioning was commissioned in 1978.

In general, ship-like vessels are noted for their mobility, high storage capacity and, usually, lower day rates. Semis are noted for their motion characteristics and are usually capable of drilling in rougher seas than the ship-like vessels. Generalities are, of course, dangerous, and the above statements apply to the extreme cases of large semis vs. conventionally designed, ship-like vessels. However, station-keeping equipment, motion-compensating equipment, the location of the rotary table with respect to the center of gravity of the vessel, and the skill and training levels of personnel all affect the performance of a vessel on the high seas. For example, turret moored or dynamically positioned vessels that can change heading to decrease vessel motion may be preferable to a small semi, if the crew and équipment of the ship are capable of superior performance.

A definite disadvantage of a semi is its limited deck load. The deck of a semisubmersible is from 40 to 70 feet above the water during drilling. This high deck, when loaded, increases the tendency of the vessel to capsize in rough seas. Also, structural integrity is a more severe problem with semis than with ship-like vessels. No ship-like drilling vessels have been lost because of structural failure, but two semis have been lost for this reason. It is hoped that the last has been seen of catastrophic structural failures. Structural integrity of semisubmersibles can be expected to improve with experience.

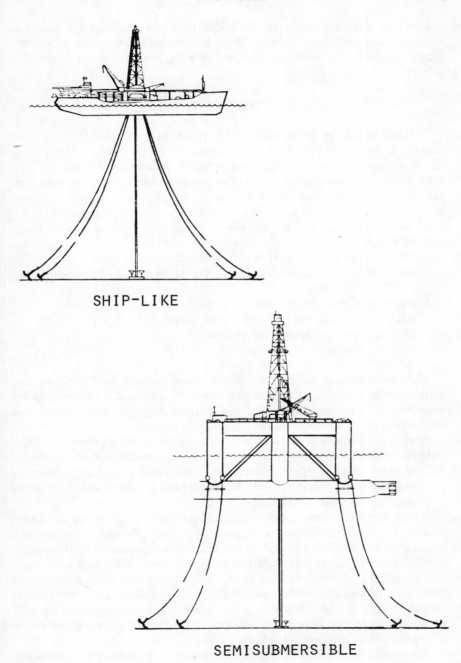

SHIP-LIKE

SEMISUBMERSIBLE

Figure 2-1. Basic types of floating drilling vessels.

Last but certainly not least, the initial cost and resulting day rates for semis are usually higher than for ship-like vessels. From the above discussion it is evident that semis are designed to work primarily in rough weather areas.

Vessel Motion

A vessel underway or moored has six motions, sometimes referred to as degrees of freedom. Described in a typical, XYZ coordinate system, three of the motions are translational and three are rotational (see Figure 2-2). It is preferable, however, to categorize these motions based on how they can be controlled. These motions are:

Motions restricted to the horizontal plane
    SURGE: Translation fore and aft (X-axis)
    SWAY: Translation port and starboard (Y-axis)
    YAW: Rotation about the Z-axis (rotation about the moonpool)

Motions that operate in vertical planes
    HEAVE: Translation up and down (Z-axis)
    ROLL: Rotation about the X-axis
    PITCH: Rotation about the Y-axis

Motions restricted to the horizontal plane can be controlled by the station-keeping system. Vessel design naturally affects the station-keeping requirements, but there is also control of these motions after the vessel is built.

Heave, pitch and roll are a direct consequence of vessel design and not much can be done to improve these motion characteristics without redesigning the vessel. Pitch and roll characteristics of a vessel can be degraded by a poor station-keeping system (see Chapter 3), but a station-keeping system will not override the basic design.

Severe pitch and roll can impede drilling, while severe heave can overtension or buckle the riser, and cause it to fail. These vessel motions increase as the "natural" period of this motion is approached by the waves. The natural periods of pitch and roll tend to decrease with increasing mass and extend along a line normal to the axis of rotation. For example, the pitch of a ship-like vessel will be greater than its roll because the vessel is longer than it is wide. Increasing the beam width will decrease the roll amplitude in choppy seas. What happens is that the natural period of roll is increased above the period of the waves.

If we build a vessel like a box and realign the mass (as on a semi), the natural periods for each motion will be altered from those of a ship. By increasing the natural periods of this vessel to values above those usually

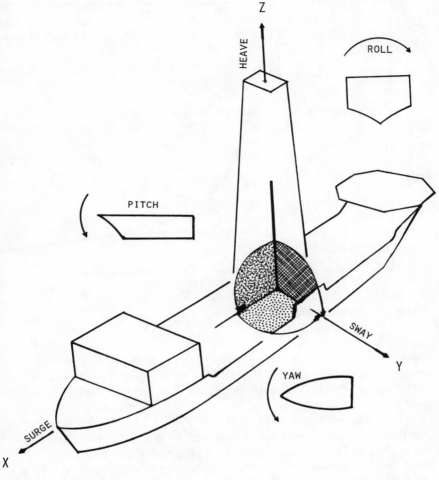

Figure 2-2. Vessel motions.

encountered in storms, we will improve the vessel motion characteristics in those particular waves. That is basically what is done with the large semis.

Comparison of Vessel Motions

To compare the advantage that one drilling vessel may have over another, their motion characteristics that may be anticipated under certain conditions must be considered.[1,2] For this discussion, two imaginary drilling vessels will be used. Call them SHIP A and SEMI B. This comparison will point out

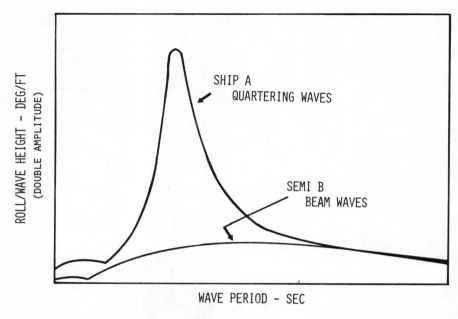

Figure 2-3. Roll in regular waves.

characteristic differences between ship-like vessels and semisubmersibles. In actual practice, this method can be used to compare *any* two vessels.

The major difference between the motions of a ship and the motions of a semi is the roll in beam waves. Waves on the quarter are bad enough, as shown in Figure 2-3. In this case SHIP A, with waves acting on the quarter, has a much lower roll period than SEMI B. Pitch is also worse for a ship than for a semi, but the pitch amplitude is much lower than the roll amplitude for a ship and is not usually a determining factor in a ship vs. semi comparison. When two semis are compared, pitch can be very important.

Heave comparisons present a somewhat different situation (see Figure 2-4). The heave of SHIP A is higher than the heave of SEMI B for periods up to the natural period for the semi. For storm waves less than this natural frequency, SEMI B may drill during the storm when SHIP A is shut down waiting on weather (WOW). But long period waves that follow a storm can force drilling operations to be suspended on SEMI B.

Figures 2-3 and 2-4 present vessel motions resulting from "regular" waves, that is, a sine wave is used as the model wave. At sea, waves are irregular, representing the sum of a series of sine waves with different amplitudes and periods (see Figure 2-5). With the sophistication of the

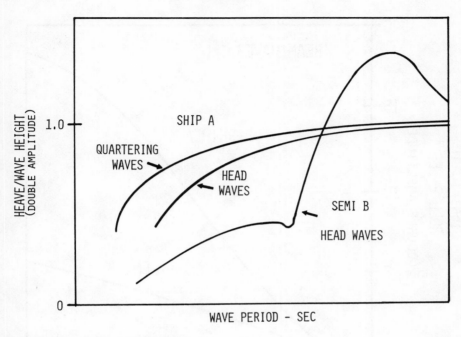

Figure 2-4. Heave in regular waves.

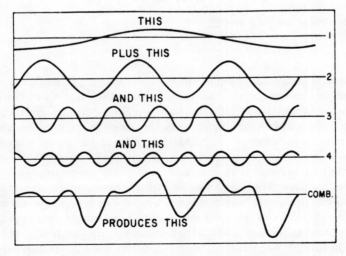

Figure 2-5. Wave pattern combining four regular waves.

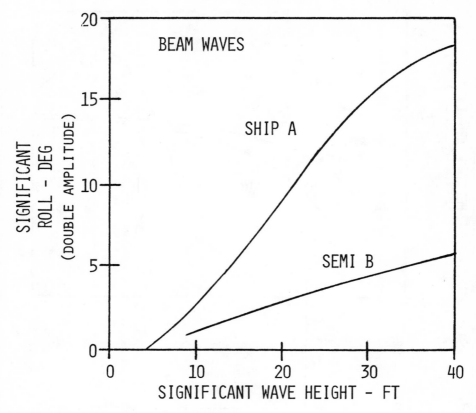

Figure 2-6. Roll in irregular waves.

model wave, a statistical comparison is required using "significant" terms. Significant in this case means an average over the largest one-third of a sample. Since irregular waves do not have a regular period, significant wave heights are plotted with significant vessel motions in Figures 2-6 and 2-7. For this particular comparison, SEMI B definitely has an advantage over SHIP A.

For ship-like vessels, both roll and heave are worse in beam waves than in head waves. Thus, one way to reduce these motions is to turn the vessel into the prevailing waves. Figure 2-8a shows how heave varies with direction for SHIP A in irregular waves. The same variations for pitch and roll are shown in Figure 2-8b. Moored vessels are limited to holding a specified heading while drilling; however, turret moored and dynamically positioned vessels can assume any heading without having to disconnect

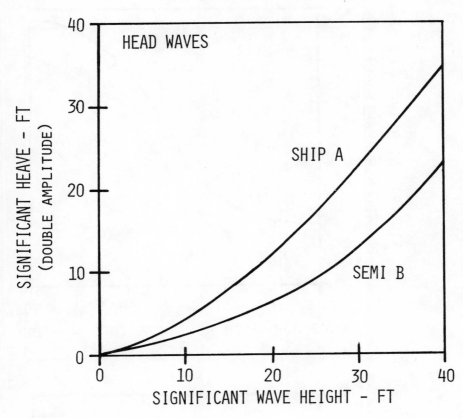

Figure 2-7. Heave in irregular waves.

from the wellhead. This maneuverability greatly improves the performance of such vessel designs.

Tolerable Vessel Motions

Now that we have some basis for estimating vessel motion as a function of significant wave height, let us consider the vessel motions that can be tolerated at different phases of a drilling program. If such data are available on a rig that is being considered, use it and consider yourself fortunate. In the absence of such data, information such as that published in *Ocean Industry*[3] can be helpful. The following data were presented at a panel discussion on vessel motion and, as might be expected, opinions varied somewhat.

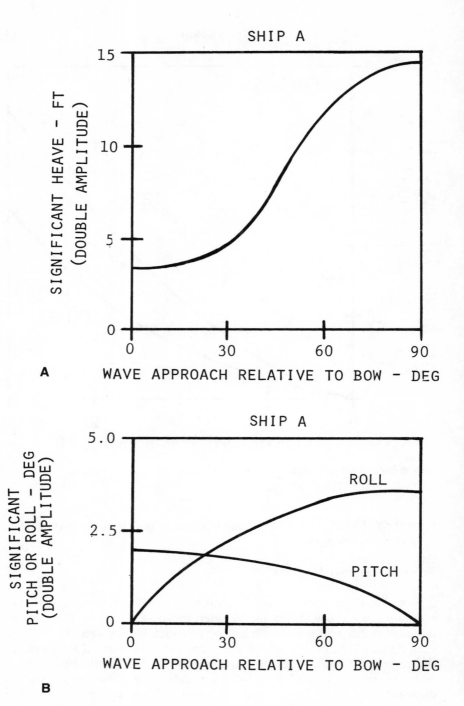

Figure 2-8. Response of ship-like vessels to wave approach angles.

Table 2-1
Motion Limits Criteria
(Wave Height and Heave)*

Operation	Wave Height (ft)	Heave (ft)
Drilling ahead	30	10
Running and setting casing	22	6
Landing BOP and riser	15	3
Transferring equipment	15	-

*Data from *Ocean Industry* Magazine.

Table 2-2
Motion Limits Criteria
(Pitch, Roll and Heave)†

Operation	Roll* (deg)	Pitch* (deg)	Heave* (ft)
Vessels with pipe handling equipment:			
Drilling and tripping	14.0	14.0	7.0
Fishing and logging	14.0	14.0	5.0
Running casing	5.0	5.0	5.0
Running BOP or riser	2.2	2.2	2.7
Vessels without pipe handling equipment:			
Drilling and tripping	6.0	6.0	7.0
Fishing and logging	9.0	9.0	5.0
Running casing	5.0	5.0	5.0
Running BOP or riser	2.2	2.2	2.7

*Significant, double amplitude motions.
†Data from *Ocean Industry* Magazine.

One company represented on the panel would discontinue operations based on wave height and vessel heave. Maximum allowable conditions for this company are shown in Table 2-1.

Another company presented limits based on maximum vessel motion that could be tolerated when doing a particular job. Their criteria are presented in Table 2-2 for rigs with and without pipe handling equipment.

These limits were believed reasonable for most vessel designs. Any vessel motions, however, must give consideration to the tolerance level of the personnel on the vessel in order to assure safe personnel performance in any particular environment.

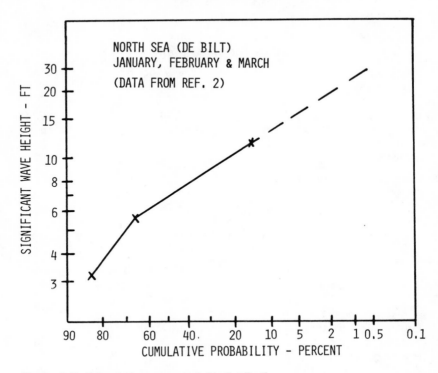

Figure 2-9. Cumulative wave height distribution.

Vessel Evaluation for a Specific Area

Now that some criteria have been established for shutting down drilling operations, this information can be used with vessel motion data and wave height distribution to compare the down time waiting on weather for our hypothetical vessels.

For example, the maximum heave to be tolerated while drilling and tripping is 7 feet, with or without pipe handling equipment. This corresponds to a significant wave height of about 14 feet for SHIP A and 21 feet for SEMI B (see Figure 2-7). Heave is the limiting factor in this case because roll and pitch for higher significant wave heights can be tolerated (see Figure 2-6). Such data can be used with a plot of cumulative wave distribution data (see Figure 2-9) to compare vessels. Cumulative wave data will be obtained for a specific area at a particular time of the year. From Figure 2-9, it can be estimated that SHIP A would be shut down 10% of the time, while SEMI B would be shut down about 1.5% of the time. This logic can be followed to develop a table for vessel motion comparisons such as Table 2-3. This table assumes that the ship will always be heading into the

Table 2-3
Comparison of Vessel Down Time for Major Operations

Operation	SHIP A			SEMI B		
	Significant Wave Ht (ft)	Down Time (%)	Limiting Factor	Significant Wave ht (ft)	Down Time (%)	Limiting Factor
Drilling and tripping	14	10	Heave	21	1.5	Heave
Logging and cementing	11	18	Heave	17	5	Heave
Running casing	11	18	Heave	17	5	Heave
Running riser and BOPS	7	50	Heave	10	22	Heave

Table 2-4
Percentage of Time Spent
on Different Operations
Example

Operation	Time (%)
Drilling and tripping	52
Logging and cementing	15
Runing casing	9
Running riser and BOPS	9
Other operations	15

prevailing waves. If the ship is subjected to beam waves, additional down time for running the casing, the riser and the BOPs may be expected.

The relative importance of each operation can be appreciated more when accompanied by a table showing the percentage of total time usually required for each operation during the drilling of a "typical" well (see Table 2-4). The percentage of any time spent on each operation will vary between different areas. Any table used should be based on data obtained from local operators or contractors.

Naturally, the experience on an individual well may differ greatly from the results long term averages might suggest. Also, the time spent waiting on weather will be influenced by many factors not included in the model that has been presented. However, calculations such as these can be useful tools for direct comparisons between different vessels and the probability of one vessel's out-performing the other.

Capacities

A floating drilling rig has capacity limitations based on both weight and volume. Load capacity refers to the maximum weight of consumable materials that can be stored on board. Stowage capacity is the space allocated for stowing these different materials. The load capacity is determined by the load line, but the vessel must be stable under loaded conditions. For example, the maximum deck load for a semi will be calculated at a specific draft. The combined draft and deck load will have a marked effect on stability.

The stowage capacity of a vessel will generally be greater than the load capacity. In other words, the vessel will probably not be able to carry safely all of the heavy iron goods, bulk mud, etc. that can be loaded on board

Table 2-5
Comparison of Vessel Sizes

Item	SHIP A	SEMI B	SEMI C
Length (ft)	400	333	295
Beam (ft)	65	318	245
Draft (ft)	21	72	80
Deck load (lt)		2,020	2,000
Dead weight (lt)	6,576	8,290	12,630
Displacement (lt)	11,065	17,325	22,300

(storage capacity is usually larger than load capacity to allow flexibility in carrying a variety of materials).

Comparing sizes and capacities of ships and semis is similar to comparing apples and oranges. Table 2-5 compares the size of a drill ship with that of two semis. The semis are heavier, and SEMI C has about twice the displacement of SHIP A. This is because the semis weigh more when empty than do ships, and they require large amounts of ballast water.

SHIP A does not have a value that is comparable to the deck load of the semisubmersibles. The dead weight given in Table 2-5 is the maximum load allowed, including all liquids. The dead weights of the semis are much higher than that of the ship; however, the dead weights include ballast water. Thus, this higher weight in the semis is not reflected in the stowage capacity (see Table 2-6). SHIP A can carry more of most of the materials listed than either of the semis. The only two exceptions are that SEMI B can carry more sacked material and potable water than SHIP A can. Logistics will obviously play a very important roll in determining any trade-offs between vessels being considered.

The major reason for the differences in capacities is that the semis carry tubular goods and bulk materials at the deck level. This deck may be located from 40 to 80 feet above the water during drilling. The liquids are usually stored in ths substructure of a semisubmersible. Ships, however, can store large amounts of tubulars and bulk sacks low in the hold. Adding weight high on a vessel decreases the stability, while adding weight low (below the center of gravity) increases the stability. Stability will be discussed later in this chapter.

The amount of equipment that can be carried aboard a vessel will determine the logistic support needed to sustain the drilling operation. The more equipment the vessel can carry, the less the logistic requirements will be. A good, but tedious, way to determine which vessel is the best for a particular operation is to estimate how much of the needed equipment can

Table 2-6
Comparison of Stowage Capacities

Item	SHIP A		SEMI B		SEMI C	
	Capacity	Where* stowed	Capacity	Where* stowed	Capacity	Where* stowed
Fuel (bbl)	8,154	below	4,830	below	6,800	below
Drill water (bbl)	15,121	below	4,830	below	11,900	below
Potable water (bbl)	513	below	3,654	below	570	below
Mud in pits (bbl)	2,920	both	2,628	on	2,000	on
Dry bulk (ft³)	16,400	below	13,280	on	9,625	on
Sacks (sk)	12,000	below	17,400	on	9,230	on
Tubular Goods (ft³)	19,000	both	7,260	on	5,000	on

*NOTE: below means below decks, on means on deck, and both means the material is stowed both above and below decks.

be loaded in each vessel. With this information, the workboat requirements, warehousing and other logistic requirements can be estimated for the operation. Then the cost of the logistics can be plugged into a mathematical model for comparison of alternatives.

Mobility

Ships have an edge over most semis where mobility is important. Calm weather speed for a ship will be from eight to 14 knots, but for most "self-propelled" semis the speed will be less than eight knots when fully loaded. However, the Aker design can make about eight knots when fully loaded. Most of the newer models of semis are towed, and use propulsion to assist in towing. The "135 series" models that have no thrusters are designed with elliptical footings and require three large tugs to make three or four knots.

Some of the semis require a reduced load carrying capacity at the designed towing draft because they are too heavy to get up to towing draft with a full load. In such cases, the load must be decreased or the vessel must be towed at a deeper draft. The deeper draft naturally increases the towing resistance. Semisubmersibles often have to be ballasted down in rough seas during a tow. This decreases the mobility especially if large horizontal cross braces are located just above the water line at towing draft.

In general, when drilling one or two wells in a remote area and then returning, a ship is preferable to a semi because of its stowage capacity and mobility. However, vessel motion down time must also be considered in the choice of a rig.

Stability

A body that floats at the surface of a liquid obeys Archimedes' principle of flotation, that is, the upward force of the water is equal to the mass of the object that is floating, and this mass is equal to the mass of water displaced. The watertight volume above the water line is called the reserve buoyancy and is important in determining the minimum service freeboard. The minimum service freeboard is a measure of the vessel's ability to survive compartment flooding and storm conditions, and is used to determine the maximum permissible vessel draft.

The *center of buoyancy* is always located at the center of mass of the displaced water. Its position depends on the shape of the vessel below the water line (WL). The vessel will always adjust itself so that the center of gravity (G) and the center of buoyancy (B) lie in the same vertical plane (see Figure 2-10).

Stability is defined as the resistance of a vessel to capsizing. It refers to the heel motion of the vessel. When the vessel heels, the center of gravity should

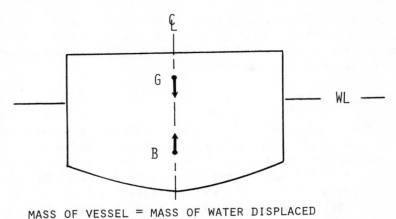

MASS OF VESSEL = MASS OF WATER DISPLACED

Figure 2-10. Schematic showing vessel at even keel.

remain in the same position relative to the vessel, but the center of buoyancy moves (see B_1, in Figure 2-11). With proper hull design, the center of buoyancy will move so that a righting moment is applied to the vessel. A vertical line through the new center of buoyancy will intersect the center line of the vessel at point M. At very small angles of heel, point M is referred to as the metacenter and the length of the line GM is called the metacentric height or simply the GM of the vessel. This height is a measure of the resistance of the vessel to heel and may also be referred to as the initial stability. The line GZ is a direct measure of the stability of the vessel at any angle of heel and is used to define the service capacity of a vessel.

The inclining of the vessel fore and aft is referred to as the trim. The trim of a vessel is especially important when underway and influences the speed and handling characteristics of the vessel. Determining vessel trim characteristics is similar to determining the stability characteristics.

To determine stability, the righting arm (GZ) of the vessel is plotted vs. the angle of heel (see Figure 2-12). The GM is the slope of the curve at zero heel angle. It is conventionally reported in feet of righting arm that the tangent makes when extended to the angle of one radian (57.3 degrees).

Now we are in a position to define the stability required by governmental authorities. These requirements are:[4,5,6]

1. *Intact stability* requires that the righting moment be adequate to withstand designed wind velocities from any direction.
2. *Damage stability* requires that a vessel must withstand flooding of any major compartment that may be reasonably expected to be flooded. The vessel under these conditions must also withstand certain wind loading from any direction.

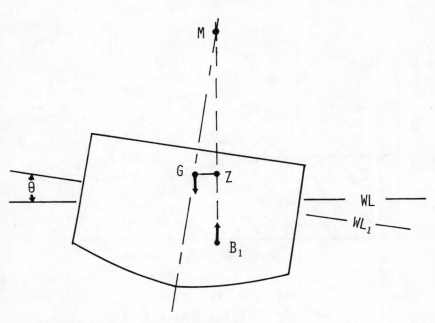

Figure 2-11. Vessel heeling at Angle Θ.

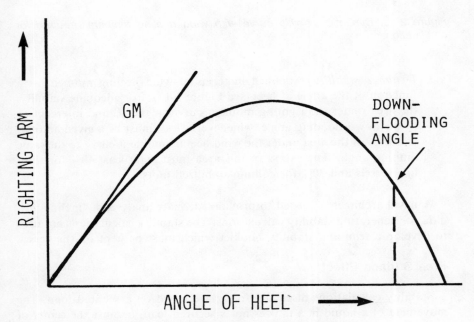

Figure 2-12. Graphical representation of righting arm and GM.

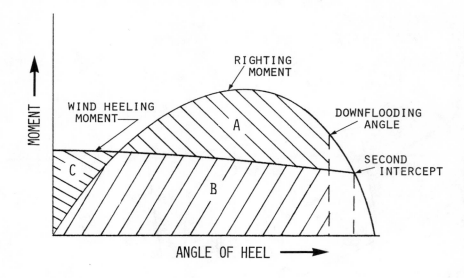

FOR SHIP-LIKE VESSELS
AREA(A + B) ≥ 1.4 AREA(B + C)
FOR COLUMN STABILIZED VESSELS
AREA(A + B) ≥ 1.3 AREA(B + C)

Figure 2-13. Dynamic stability curve with wind heeling moment required for certification.

3. *Dynamic stability* is defined in terms of wind heeling moment as a function of the angle of heel (see Figure 2-13). For adequate stability, the area under the righting moment curve to the second intercept or to the down-flooding angle, whichever is less, must be a given amount in excess of the area under the wind heeling moment curve to the same limiting angle. The excess of this area must be at least 40% for ship-like vessels and 30% for column-stabilized units.

A naval architect is needed to provide adequate analysis of "incline-test" data for generating stability calculations. The stability calculations are used to develop a trim and stability booklet which must be kept on the vessel.

Free Surface Effect

Stability calculations discussed so far are based on a secured load. The movement of a liquid in a tank is not a secured load, because the center of

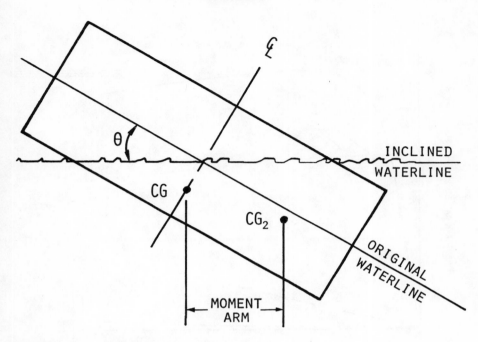

Figure 2-14. Free surface effects.

gravity of the tank changes with vessel motion. Seagoing vessels are designed with partitioned tanks that will not present a stability problem unless they are cross-connected. Dangerous situations can occur because of the free surface phenomenon, and personnel who ballast vessels and transfer liquids between tanks should have an understanding of problems that can be caused by partially filled tanks.

If a tank is half full of liquid and inclined at an angle θ (see Figure 2-14), the center of gravity will move from position CG to some other position GG_2. The moment change caused by this change in the center of gravity is referred to as the *moment of transference*. The moment of transference is appreciably affected by the lateral extent of the tank and, to a lesser extent, by the height.

To illustrate this, consider three tanks with equal volume and cross section but with varying heights (d) and lateral extents (L). All three tanks will be half full of liquid so that the liquid height (h) equals ½d. Changes in moment arm with lateral inclination of the tanks are shown in Figure 2-15. The moment arm is proportional to the moment of transference because all tanks contain the same volume of liquid with the same density. The higher values are for the 21-meter-wide tank, and the curve goes through

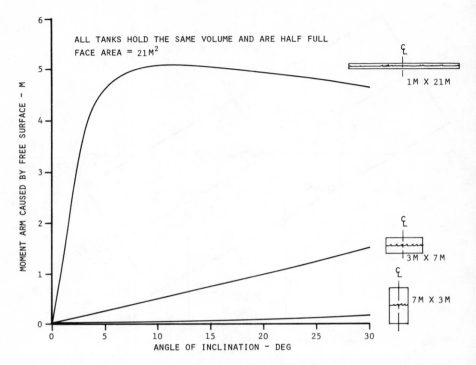

Figure 2-15. Effect of tank geometry on the moment of transference of half-filled tank.

a maximum that is the result of the liquid's filling the lower end of the inclined tank. The three-meter-wide tank has the lowest moment of transference of the tanks shown in the figure.

Now let us vary the amount of liquid in the 21-meter-wide tank to see how fullness affects the moment of transference (see Figure 2-16). The upper set of curves is the moment of transference and is more important than the lower set that shows how the moment arm increases with decreasing fullness of the tank. Comparison of the two sets of curves shows how the moment arm and the mass of the liquid combine to produce the maximum moment of transference with the half filled tank. In this example, the tank has a volumetric capacity of 840 cubic meters, and pure water (SG = 1.00) was used. Thus the tank will hold 840 metric tons.

Cross-connecting wing tanks. When wing tanks in a ship are cross-connected, the moment of transference is increased by liquids moving from one wing to the other. The results are a loss in stability, an increase in roll, and a list to one side. The rate of stability loss depends on the flow resistance

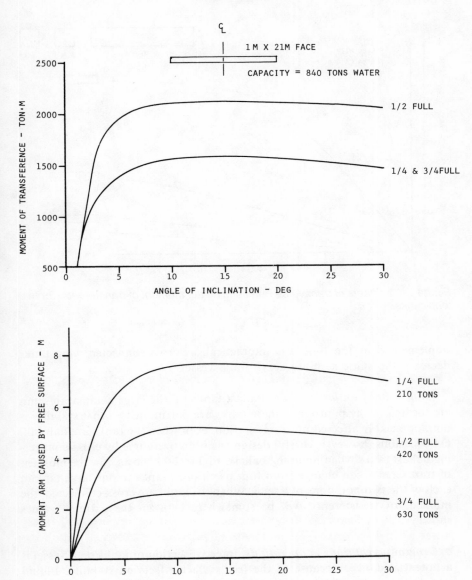

Figure 2-16. The half-filled tank is the worst case.

of the lines that connect the tanks. Cross-connected wing tanks present a hazardous situation, especially in workboats (see Chapter 9).

At one time, the physics of cross-connecting wing tanks was considered for anti-roll stabilization of drilling vessels. The way that this idea was

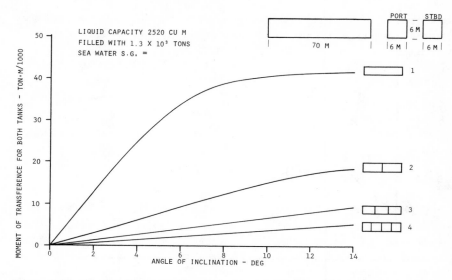

Figure 2-17. Effects of transverse partitions on the moment of transference for two long tanks.

implemented at the time was impractical, as cross-connected wing tanks decrease the static stability curve for a vessel.

Ballast tanks on large semis. Ballast tanks on the large catamaran semis are located in the pontoons. These tanks are partitioned so that stability is not appreciably affected by vessel motion, even when the tanks are half full. On occasions, however, careful design has been overcome by cross-connecting ballast tanks when pumping ballast. Figure 2-17 shows how the moment of transference can change from four partitioned tanks to one large tank in each of the two pontoons of a semi. When the tanks are cross-connected, the moment of transference will be somewhere between the extreme values shown.

General precautions in operations. Interconnection of tanks on a vessel is a potential danger because of the free surface effects on stability. Liquid transfer lines should be closed except when transfer of liquids is required. Transfers should be accompanied by controlled ballasting of the vessel.

Classification and Certification

No classification society either writes or enforces any rules. Classification societies do, however, issue certificates testifying that rules of certain government bodies have been complied with. Three classification societies are particularly important to offshore drilling. These societies are:

CLASSIFICATION SOCIETY	COUNTRY	GOVERNMENTAL AUTHORITY
American Bureau of Shipping	U.S.A.	U.S. Coast Guard
Det Norske Veritas	Norway	Norwegian Petroleum Directorate
Lloyd's Registry of Shipping	U.K.	British Admiralty

The differences in the requirements of the different countries are only minor, but they can get you into trouble. Countries other than the three mentioned usually specify which rules of which of these three countries apply. If a vessel is to be built, the building codes should be studied and the most stringent code should be considered so as not to limit potential operating area.

Sometimes it is advisable to have a vessel "built under survey." This way, surveyors from the classification society observe the construction to make sure that a vessel is built according to the rules. The society charges for this service, and issues a certificate that the vessel was built under survey. Building a vessel under survey is not a legal requirement, but it does assure initial certification.

Drilling vessels not built under survey may receive classification by passing a certification survey. The approach used depends on the owner.

Periodic surveys are required, and additional, unscheduled, surveys may be required as the vessel moves from country to country. Keeping up with the different requirements avoids being shut down by governmental authorities.

Termination of Classification

Continuation of the classification of any seaworthy unit is contingent upon:

1. Periodic inspections specified in the classification certificate.
2. Surveys after damage and repair.
3. Authorization for and survey after any modification to the vessel.

Surveyors representing the societies listed previously are located around the world and can be contacted with relative ease. The captain (or barge-master) of a vessel should be familiar with the requirements of the authority for classification, including what structural changes are classified as modifications. For example, welding of the outer hull including the decks can be considered a modification. Such modification without proper authority and inspection can lead to rig shut down.

References

1. Stanton, P.N., and Kuang, J.G. "Evaluation of Semisubmersible Drilling Vessels," OTC paper 2031. Presented at the Offshore Technology Conference, Houston, 1974.
2. Michel, W.H. "How to Calculate Wave Forces and their Effects," *Ocean Industry* (May 1967).
3. "Panel Discusses Offshore Drilling Technology," *Ocean Industry*, May 1972.
4. *Rules for Building and Classing Offshore Mobile Drilling Units*, American Bureau of Shipping, New York, 1973.
5. *Rules and Regulations for Construction and Classification of Steel Ships*, Lloyd's Registry of Shipping, London, 1974.
6. *Regulations for Mobile Drilling Platforms*, Norwegian Maritime Directorate, Oslo, July 11, 1969, as amended Sept. 10, 1973.
7. "Weather Averages and Sea States for Selected Offshore Areas," *Ocean Industry* (Jan. 1973).

Note: Tables of weather data were published in *Ocean Industry* (from about April 1971 through July 1973).

Station Keeping

3

The ability of a vessel to maintain position for drilling determines the useful time that a vessel can effectively operate. Stated negatively, if the vessel cannot stay close enough over the well to drill, what good is the drilling equipment? Station keeping equipment influences the vessel motions in the horizontal plane. These motions are: surge, sway and yaw. Generally, surge and sway are the motions that are considered. Yaw motion is decreased by the mooring system but is neglected in most mooring calculations.

When investigating or designing a mooring system, the following criteria should be considered:

1. *Operational*: The vessel is close enough over the well for drilling operations to be carried out. This varies between operators, but is usually 5% or 6% of water depth. Later, other criteria, based on riser considerations, will be discussed.

2. *Non-operational but connected*: The condition from the operational stage up to 10% of water depth. Drilling operations have been stopped, but the riser is still connected to the wellhead and BOPs.

3. *Disconnected*: The riser is disconnected from the wellhead and the BOPs, and the vessel can be headed into the seas.

The operating stage (*stage 1* above) is the primary criterion for evaluating a station keeping system. *Stage 2* is needed for estimating maximum line tensions and minimum line lengths.

Logic similar to the logic used to compare vessel motions may also be used to evaluate or compare station keeping systems. Namely, what percentage of the time will a rig be able to hold station so that it can drill? This requires comparing the mooring or thruster restoring forces with the environmental forces that are anticipated. The probability of wind, wave and current forces must be considered for judging if a ship is suitable for drilling at a given location.

Methods for estimating the effects of the different environmental forces are included in the mooring section. These estimates are sometimes less accurate than we would like, but they have been adequate for evaluating conventional mooring systems. The same techniques may be used for dynamically positioned drilling rigs. Extensive model test data, however, are required for dynamically positioned drilling (not coring) rigs and should be available for vessel evaluation.

Mooring

Calculating wind and current forces on a vessel are reasonably straightforward processes; however, wave data are more difficult to evaluate. Waves load the mooring lines through short variations in vessel position, and add to the steady state vessel offset. The tension fluctuations that accompany position oscillations are important because they cause periodic loading of the mooring lines. In storms, these fluctuations in tension may be appreciable: they can cause lines to break and anchors to drag. To estimate this maximum line loading, vessel response to waves will be divided into two catagories:

1. *Long term drift.* Considered a steady state force, and added to the forces of wind and current to calculate the total force applied to the vessel.

2. *Short term motions.* Fluctuations about a mean offset. These motions are caused by waves as they slap against the hull. For any vessel, these fluctuations are considered to depend on significant wave height only and are independent of offset.

Both motions will effect the tensions in the mooring lines. The application of these motions to calculate offset and tension will be discussed later in the text.

Environmental Forces

To calculate the environmental forces on a moored vessel, we must have an estimate of the wind, current and waves that will be pushing on the vessel. These environmental conditions should be expressed in terms of the probability of occurrence. All forces should be assumed to be coming from the same direction, unless an alternate criterion can be justified. The environmental data are converted to forces and balanced against the restoring forces available from the mooring system to estimate the amount of rig down time that can be expected. Mooring systems have operational

Table 3-1
Shape Coefficients[1]

Shape	C_s
Cylindrical shapes	0.5
Hull (surface type)	1.0
Deck house	1.0
Isolated structural shapes (cranes, beams, etc.)	1.5
Under deck areas (smooth surfaces)	1.0
Under deck areas (exposed beams and girders)	1.3
Rig derrick (each face)	1.25

flexibility which enables them to meet different situations without making major modifications to the mooring system.

Wind force. The following equation is specified by the American Bureau Shipping (ABS)[1] and is internationally accepted:

$$F_A = 0.00338 V_A^2 \cdot C_h \cdot C_s \cdot A$$

where:

F_A = wind force, lb
V_A = wind velocity, knots
C_s = shape coefficient from Table 3-1, dimensionless
C_h = height coefficient from Table 3-2, dimensionless
A = projected area of all exposed surfaces, ft.[2] This area changes with both heel and yaw.

Recommendations to expedite combining areas for calculation are included in reference one. In brief, adjacent areas of like shape and height coeffecients can be grouped. These area groups are multiplied by their pertinent coefficients and summed to give the wind force as a function of wind velocity.

Current force. The following equation can be used to calculate current drag forces:

Table 3-2
Height Coefficients[1]

Height Design water level to center of area		C_h
from–to (m)	from–to (ft)	
0– 15.3	0– 50	1.00
15.3– 30.5	50–100	1.10
30.5– 46.0	100–150	1.20
46.0– 61.0	150–200	1.30
61.0– 76.0	200–250	1.37
76.0– 91.5	250–300	1.43
91.5–106.5	300–350	1.48
106.5–122.0	350–400	1.52
122.0–137.0	400–450	1.56
137.0–152.5	450–500	1.60

where: $F_c = g_c C_s V_c^2 A$

F_c = current drag force, lb
C_s = drag coefficient, dimensionless. Same as the wind coefficient (Table 3-1)
V_c = current velocity, ft/sec
A = projected area, ft^2
$g_c = 1 \left(\dfrac{\text{lbft·sec}^2}{\text{ft}^4} \right)$

The density of water is assumed to be 64.4 lb/ft^3

Note: The above formula differs from that in reference 1 because the drag force and areas are calculated with respect to total area and total force. The references calculations are based on force per unit length.

This equation can be used for drag forces on mooring lines, risers and other objects. To be accurate, the current forces acting on the riser, mooring lines and other vessel appendages should be included. Experience indicates, however, that for operations noted by the author, forces of the waves and current acting on the mooring lines and riser are negligible compared with the forces operating on the vessel.

Wave force. For semisubmersibles, the classification societies will consider any valid theoretical approach or alternate test data for determining the effects of waves. With the lack of these data, the ABS recommends the use of a shallow water theory based on an interpolation between the values of the solitary wave theory[2] and the Airy wave theory[3]. For deep water, a sine wave theory is accepted by the ABS. These calculations are detailed in reference 1. The newer semis have wave tank test data that are more reliable and should be used.

For a ship-like vessel, the best source of wave force data are also from model tests. In the event that model test information is not available, the following equations can be used:[4]

Bow forces:

for $T > 0.332\sqrt{L}$

$$F_{bow} = \frac{0.273\ H^2 B^2 L}{T^4}$$

for $T < 0.332\sqrt{L}$

$$F_{bow} = \frac{0.273\ H^2 B^2 L}{(0.664\ \sqrt{L} - T)^4}$$

Beam forces:

for $T > 0.642\sqrt{B + 2D}$

$$F_{beam} = \frac{2.10\ H^2 B^2 L}{T^4}$$

for $T < 0.642\sqrt{B + 2D}$

$$F_{beam} = \frac{2.10\ H^2 B^2 L}{(1.28\sqrt{B + 2D} - T)^4}$$

where:

T = wave period, sec
F = wave force, lb
H = significant wave height, ft
L = vessel length, ft
B = vessel beam length, ft
D = vessel draft, ft

Short term motions from waves. Short term vessel motions are small oscillations about a mean offset. The mean offset is established by summing

Table 3-3
Gross Estimates of Short Term Motions
of Moored Vessels

Vessel Type	Percentage of Wave Height*	
	Surge	Sway
Wave periods less than 10 sec		
Ship-like	25	50
Semi (symmetrical)	25	25
Semi (catamaran)	25	40
Wave periods greater than 10 sec		
Ship-like	50	50
Semi (symmetrical)	38	38
Semi (catamaran)	38	50

*This table uses the convention that:
  1. Wave height is double amplitude, measured crest to trough.
  2. Surge and sway are single amplitude values.

the wind, current and wave (long term) forces that are in equilibrium with
the restoring forces of the mooring system. The short term motions should
be determined from model tests in a wave tank. In the absence of such data,
Table 3-3 may be used. These data are gross approximations based on
limited information and are not recommended if more relevant data are
available.

Use of this short term wave data will be explained later and in more detail
in the section on operational considerations.

Mooring System

The mooring system acts as a spring to resist the offsetting of the vessel by
environmental forces. Just like a spring, the restoring force increases with
increasing offset. The rate at which this force increases is conventionally
referred to as the hardness or stiffness of a mooring system. Mooring
calculations are used to determine how well a given mooring system will
function under given environmental conditions. These calculations will be
used to judge the initial tension that should be used on the anchor line when
the vessel is directly over the well, and the maximum tension to which the
anchors should be set to avoid dragging an anchor under storm conditions.
These calculations combined with environmental data will lead to an
estimate of how much time the rig will be operational (within the maximum
offset for drilling) mode, in the non-operational mode (from the operational

mode up to 10% offset), and in the disconnected mode (beyond 10% of water depth).

The restoring force calculations are made using the catenary equation. This basic equation has been used for years to design suspension bridges, and is discussed in mechanical engineering and mathematical text books.[5] The boundary conditions, however, differ between those of mooring lines and those included in most texts. The calculations can be speeded up somewhat by applying the method of Adams,[6] using a dimensional technique for estimating offsets and line tensions. Although the theory is relatively simple, calculating the total forces on a moored vessel is tedious, as will become evident during the discussion to follow.

The catenary equation describes a line that is suspended at its two ends and allowed to sag under its own weight. By changing the boundary conditions from the fixed catenary, a mooring line can be represented mathematically. Figure 3-1 shows a diagram of a mooring line. The general catenary equation for mooring use is:

$$y = \frac{H}{w} \cosh\left(\frac{xw}{H}\right)$$

and the equations used for mooring calculations for one single weight line are:

$$H = T - wd = T \cos\theta$$
$$\theta = \cos^{-1}(H/T)$$
$$V = \sqrt{T^2 - H^2} = H \tan\theta$$
$$s = \frac{V}{w} = \frac{H}{w} \tan\theta$$
$$x = \frac{H}{w} \ln\left(\frac{T+V}{H}\right) = \frac{H}{w} \ln(\sec\theta + \tan\theta)$$
$$L = x + A - s$$

where:

T = tension of the line, lb
θ = angle of the line with respect to the horizontal, degrees
H = horizontal restoring force, lb. H is constant over the length of the suspended line for any given value of T.
w = line weight per unit length, lb/ft
s = suspended line length, ft
d = water depth (should include height of outboard fairleader above water line), ft

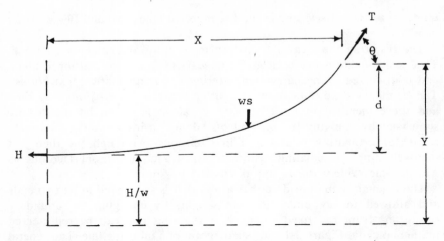

Figure 3-1. The catenary as used for mooring calculations.

y = ordinate = $d + H/w$, ft

x = horizontal distance from the vessel to the point where the line touches the seabed, ft

H/w = a translational boundary condition used to account for the force H, ft

L = horizontal distance from the vessel to the anchor, ft

A = total mooring line length, ft

The length of line, A, is not critical to the calculation but it is critical to operations. Line lengths of five times water depth in water less than 600 feet and about three times water depth in deep water are used. The minimum line length to avoid a vertical force on the anchor should be calculated at the maximum offset anticipated.

To calculate T and H vs offset:
1. Calculate initial conditions using a calm water tension
   a. Choose a calm water tension, T_0
   b. Calculate H_0
   c. Calculate L_0
2. Choose a tension T_1 and repeat calculations for H_1 and L_1
3. Calculate the offset
$$\text{offset} = L_1 - L_0 \text{ ft}$$
4. Choose tensions T_i and calculate corresponding values of H_i and offset. Continue until the offset is greater than the anticipated maximum

Table 3-4
Example of Single Line Restoring Forces

Offset % (of water Depth)	Tension (kip)	Line Angle (deg)	Single line Restoring Force (kip)
10	121.2	39	94.3
8	106.1	42	79.3
6	94.1	44	67.3
4	84.4	47	57.6
2	76.6	50	49.7
0	70.0	52	43.2
-2	64.5	54	37.7
-4	59.9	56	33.0
-6	55.9	59	29.0
-8	52.5	61	25.6
-10	49.6	63	22.7

Water depth = 400 ft
Chain size = 2½ in.
Chain weight = 66.8 lb/ft (in water)

The above calculations are for just one line of a constant weight. Composite lines (wire rope and chain) require additional iterations to determine the horizontal force at a given tension.

The restoring force of the entire system is the vector sum of contributions of all of the lines, depending on the vessel offset. This is obviously a tedious hand calculation because of the bookkeeping required. However, it is easily solved on a computer.

Using a computer program is always advisable, but we can get a feel for how a mooring system works based on calculations simpler than those for an entire mooring system. For example, consider a linear system where two lines are directly opposed and acting respectively on the bow and stern of a vessel. Table 3-4 shows the line tension, restoring forces and line angle as a function of vessel offset for such a system.

If the weather is calm and there is no current, the vessel offset is zero and both lines are exerting 43.2 kips in opposite directions. If a force (G) pushes the vessel to a 4% offset, then the windward line is resisting G with 57.6 kips but the leeward line is assisting G with 33.0 kips, and G = 57.6 - 33.0 = 24.6 kips. Operationally, the vessel can be adjusted by slacking the leeward lines slightly and redistributing the tension. Just a few feet of slack is required.

Suppose now that the storm continues to build. The vessel offset is above 6% and seems to be increasing rapidly. We need more restoring force. If the leeward line were slacked off to a 90° angle, the resultant restoring forces would be shown as follows:

Offset (Percent of water depth)	Restoring Forces (kip)	
	Both Lines Tensioned	Leeward Line Slacked to 90° angle
6	38.3	67.3
8	53.7	79.3
10	71.6	94.3

As in this case, slacking the leeward lines on a vessel will increase the mooring line restoring forces. In operations it is not advisable to slack off to a 90° angle because the vessel may yaw excessively if the lines are too slack on the leeward side.

This is one example of how a simple model can be used to understand the operation of a mooring system. Other models can be used to understand better the relative hardness of a mooring system.

System hardness. Since the mooring system acts as a spring to resist environmental forces that try to push the vessel off from the drilling location, the mooring must be hard enough to hold the vessel in position. On the other hand, the mooring must not be so hard that response to short term forces cause the vessel to shudder as a small dog suddenly jerked back by an unyielding leash. Too hard a mooring system will result in undesirable vessel motions and excessive line fatigue.[7]

The hardness of a mooring system depends on line weight, "calm water" or "initial" tension, and water depth. The most direct method of determining hardness is the catenary angle. The smaller the angle, the harder the system. Another indication of the hardness, however, is demonstrated in the slope of the tension vs offset curves. The examples directly following are merely to show the effect of line weight, calm water tension and water depth. In each figure, the restoring forces will go from the sublime to the ridiculous and are not intended to represent any one operational situation.

Figure 3-2 shows the effect of line weight for a single line in 500 feet of water with 30 kips initial tension. Under these conditions, the mooring would be too hard with 10 lb/ft lines. A 20 lb/ft system is still too hard, but could be softened by adding chain. Additional calculations would be required to determine how much chain. The 30 lb/ft line looks good while heavier lines look too soft at this water depth and initial tension.

The hardness, however, can be increased by increasing the initial tension on a given line at the same water depth. Figure 3-3 shows what latitude exists

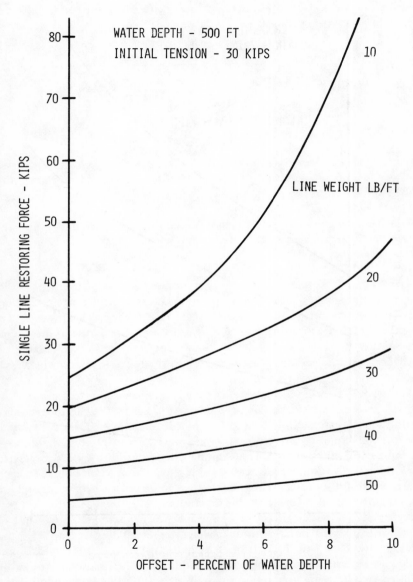

Figure 3-2. The effect of changing line weight—single-line calculations.

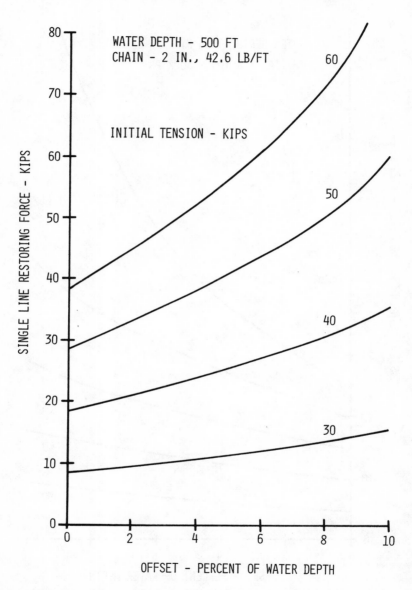

EFFECT OF INITIAL TENSION

WATER DEPTH – 500 FT
CHAIN – 2 IN., 42.6 LB/FT

INITIAL TENSION – KIPS

SINGLE LINE RESTORING FORCE – KIPS

OFFSET – PERCENT OF WATER DEPTH

Figure 3-3. The effect of changing initial tension only—single-line calculations.

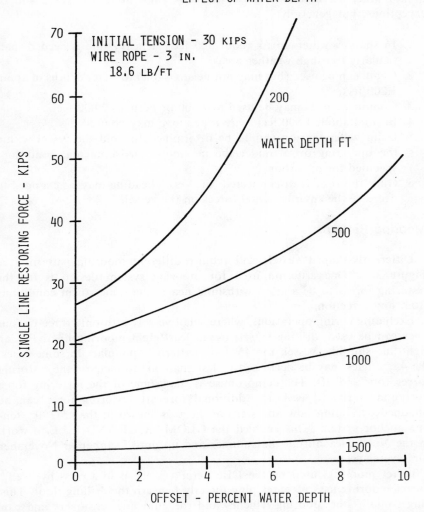

Figure 3-4. The effect of changing water depth only—single-line calculations.

in this particular system. The range in choice of initial tension will be determined by the restoring force required.

The hardness of a mooring system decreases with water depth as shown in Figure 3-4.

Hopefully, the above discussion has led to a better understanding of catenary mooring systems and will aid understanding of some of the rules of thumb practiced in industry. As with any rules of thumb there will be exceptions, but generally:

1. In shallow water up to about 500 feet, a heavy line is needed, particularly in rough weather areas.
2. Chain can be used (but may not be advisable) to water depths of about 1,200 feet.
3. Composite lines may be used to 1,500 or perhaps 2,000 feet.
4. Beyond about 1,800 feet, wire rope alone may be used.
5. Calm water tension should be determined to hold the vessel within the operating offset under the maximum environmental conditions specified for operation.
6. Once the riser is disconnected, the vessel heading may be changed to decrease the environmental forces on the vessel.

Mooring Patterns

Differently-shaped vessels will require different mooring patterns (see Figure 3-5). One criterion used for mooring system design is for the restoring forces to be able to withstand nearly the same storm conditions from any direction.

Excluding coring operations, where small vessels with only three or four lines can be used, drilling vessels use at least eight mooring lines. For an eight-line ship-like vessel, the 45°/90° pattern is popular. In some cases, the 45° angles may be decreased to less than 30° to increase the restoring forces fore and aft. The compromise is that some of the restoring force port and starboard is lost. Additional fore/aft restoring force can be obtained by adding bow and stern anchors as shown in the 45°/90°, ten-line anchor system. This enabled the GLOMAR GRAND ISLE to work in the North Sea during the summer and up until October or November for several years.

Turret mooring, used by the Discoverer vessels, is in a class by itself.[8] Turret mooring uses a center plug directly beneath the drilling floor. This plug contains the mooring winches and the guideline tensioners and can remain stationary relative to the ocean floor while the vessel rotates. An eight-line, symmetrical pattern is used and the vessel, aided by fore and aft thrusters, can be turned to keep the bow into the prevailing environmental forces. These vessels vary from the original Discoverer class (about 350 feet to the Discoverer 534 and the Seven Seas that are 534 feet. The Discoverer 534 is designed with a mooring system for over 3,000 feet of water and is dynamically positioned (see Figure 3-6). The more conventional vessels are the vessels with a spread mooring pattern. Let us consider a conventionally-moored vessel that normally uses an eight-line, 45°/90° pattern as shown in Figure 3-7.

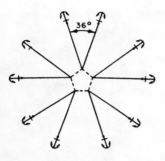

SYMMETRICAL, TEN-LINE

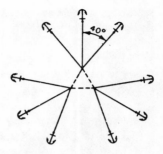

SYMMERTRICAL, NINE-LINE

TYPICAL MOORING PATTERNS FOR NON-RECTANGULAR SEMIS

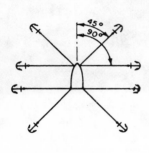

45°/90° EIGHT-LINE

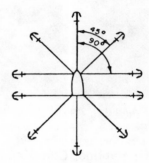

45°/90° TEN-LINE

**TYPICAL MOORING PATTERNS FOR SHIP-LIKE VESSELS
AND RECTANGULAR SEMIS**

Figure 3-5. Examples of conventional mooring arrangements.

Table 3-5 shows how the restoring forces of the mooring system will vary with offsetting forces from zero to 90° when line angles are varied. For this table, lines 1, 2, 7 and 8 were varied from 20° to 60° with respect to the center line of the vessel (angle a). The angles of lines 3 through 6 (breast lines) were held constant at 90° for simplicity. From Table 3-5, it is evident that increasing angle a will increase the lateral restoring forces at the expense of the fore/aft restoring forces. This is true as long as angle a is changed symmetrically on all four lines as shown. Varying the breast lines from 90° will increase the fore/aft restoring forces at the expense of the port/starboard restoring forces.

Figure 3-6. A fishes'-eye-view of turret mooring—Discoverer 534.

Operational Considerations

Suppose we have been able to satisfy ourselves that the mooring system discussed in the previous section is adequate as far as wind, current and long term wave forces are concerned. Also assume that the 40°/90° pattern is to be used. From the vessel motion considerations discussed in Chapter 2, it is estimated that the vessel can drill in waves that will cause a short term surge or sway of about ±3 feet, or slightly over ½% of water depth. So, drilling will have to be restricted to a mean offset of about 4½% of water depth for the worst case conditions. When disconnecting, the maximum offset should be 10% of water depth. For this vessel, the short term motion based on the anticipated waves will be about 9 feet. This is about 2% of water depth, and means that the vessel will oscillate between 6% and 10% of water depth. Thus, the riser should be disconnected either before or when the mean offset reaches 8% of water depth.

The maximum tension anticipated is 168 kips and is well below one-third of the breaking strength of 1,045 kips. The line available is a problem, however. At 10% of water depth, the suspended line length is about 1,385 feet and leaves only 115 feet of line on the ocean floor. Suppose that the riser

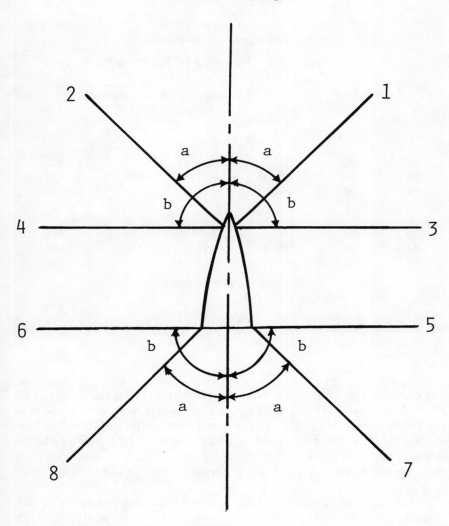

LINE—3" CHAIN, FLASH WELDED
WEIGHT—77.6 LB/FT (IN WATER)
LINE BREAKING STRENGTH—1045 KIPS
CHAIN AVAILABLE—1500 FT/LINE
WATER DEPTH—500 FT
MAXIMUM OPERATING OFFSET—5% OF WATER DEPTH
HANG-OFF BY—10% OF WATER DEPTH

Figure 3-7. Typical eight-line mooring pattern.

Table 3-5
Effects of Mooring Line Patterns

| Line Angles (deg) | | Mooring Line Restoring Force (kip) | | | |
| a | b | Displacing Force Direction | | | |
		0°	30°	60°	90°
		Offset—5% of water depth			
20	90	81	85	95	100
40	90	57	73	108	124
60	90	29	59	121	152
		Offset—6% of water depth			
20	90	99	103	115	121
40	90	69	89	130	151
60	90	35	72	147	184
		Conditions at 10% offset			
40	90	121	160	236	274

Maximum tension = 168 kips
Suspended line length = 1,385 ft

was disconnecting in time, but an unexpected surge carried the vessel to 12% offset. Now a vertical force of 16 kips will be applied to an anchor and may cause it to drag. Dragging an anchor could damage the wellhead. The least of the problems that might result would be the need to regain location, a situation that may cause appreciable down time. Because of this, additional chain should be considered, based on an evaluation of the risks and costs involved.

The above values and criteria for evaluating a mooring system are typical, though some operators may prefer a less conservative approach. Such evaluations are influenced strongly by the person or persons making the investigation.

Calculator for Mooring Data

Various programs for calculating mooring-line restoring forces and other important mooring data are commercially available. The most popular program at present is CALMS, developed and sold by Exxon Production Research Company. This program is also available on some of the time-sharing terminals in the United States.

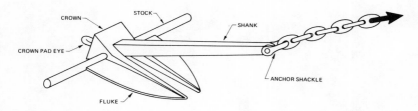

Figure 3-8. Drag anchor nomenclature.

This author has developed a mooring program for use on a portable calculator (HP-97). The advantage of this program is that it can be carried with the calculator and used in remote locations, even on the rig.

There are other programs available. All programs should have the following capabilities:

1. Be able to calculate the total restoring force and tension in the most loaded line vs. offset.
2. Be able to handle a minimum of ten mooring lines.
3. Be able to handle composite line data for wire rope and chain.
4. Have iteration limits such that the worst error for calculating forces in a line will be less than 0.1% of the smallest value anticipated.
5. Include stretch in both the wire rope and the chain. Errors of over 30% have been encountered when chain stretch was not included.

Anchors

Drag anchors (dynamic anchors) are almost universally used to hold floating drilling vessels on location. Figure 3-8 shows the nomenclature conveniently used for drag anchors. Conventional achors in use for floating drilling are the Navy LWT (Figure 3-9), Danforth (Figure 3-10) and STATO (Figure 3-11).

Sea floor conditions severely affect the anchor holding capability, and the following rules of thumb[4] are suggested for fluke angles depending on seabed conditions. These are about 50° for soft seabeds, 30° to 35° for sand and 35° to 40° for seabeds of unknown soil characteristics.

Most, if not all, of the anchors are deployed by workboats or anchor boats. The workboat tows the anchor to the setting point using the pendant line (a line attached to the crown pad eye). The mooring line attached to the anchor is payed out from the drilling vessel. The boat is maneuvered to the correct position relative to the ship. Then the anchor is lowered slowly to the seabed with the drilling rig holding a slight tension on the anchor line. This

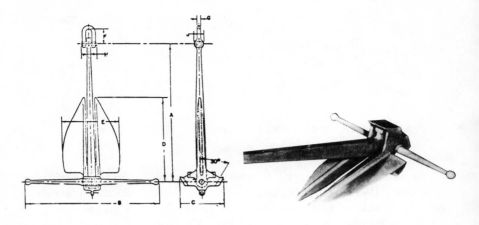

ANCHOR WEIGHT	A	B	C	D	E	F	G	H	J	PROOF TEST LBS
100	41 1/2	39 1/2	12 3/16	24	15 5/8	3 3/4	7/8	3 3/8	2 1/4	2,600
150	46 3/4	44 1/2	13 11/16	27	17 3/8	5 1/2	1	4	2 3/8	4,200
200	47 1/2	45	14 3/16	29	18 7/8	5 1/2	1	4	3 1/8	5,200
250	47 1/2	45	14 9/16	29	18 7/8	5 1/2	1	4	3 1/8	6,275
300	53	50	16 3/16	32 1/2	21 1/4	7	1 1/4	5	3 1/4	7,350
350	53	50	16 11/16	32 1/2	21 1/4	7	1 1/4	5	3 1/4	8,325
400	57 1/2	54 1/4	17 3/4	35 1/4	23	7	1 1/4	5	3 1/2	9,300
450	60	56 1/2	18 3/8	36 3/4	24	7	1 1/4	5	3 3/4	10,275
500	61 1/2	58 1/2	19	37 1/2	24	7 1/2	1 1/2	6	4 1/4	11,250
750	69	64 1/2	2 1/8	42	28 7/16	7 1/2	1 1/2	6	4 7/8	15,500
1,000	75	71	24 1/2	46	29 1/4	9 3/4	2	8	5 1/4	18,800
2,000	92 1/2	85	30	56 1/2	37 1/4	12	2 1/2	9 1/2	7	32,500
3,000	108 1/2	104	34 3/4	66	40 3/8	15	3	11	7 3/4	45,000
4,000	116	110	37 1/2	71	44	15	3	11	9	57,500
5,000	118	112	38 3/4	72	44 3/4	17	3 1/2	12 1/2	9 1/2	69,500
6,000	124	118	41	76	47 1/4	19	4	14	10 1/4	81,500
7,000	124	118	41	76	47 1/4	19	4	14	10 1/4	92,375
8,000	128	121	43 1/2	78	50 1/4	19	4	14	12	103,250
9,000	133	126	45 1/4	81 1/8	52 1/4	19	4	14	12 1/2	114,125
10,000	144	137	49	88	55 1/4	22	4 1/2	16	13	125,000
11,000	144	137	49	88	55 1/4	22	4 1/2	16	13	135,665
12,000	146	138 1/2	49 3/4	89 1/8	57 3/8	22	4 1/2	16	13 3/4	147,335
13,000	154	146	52 1/2	94	60 1/2	24	5	17 1/2	14 1/2	157,000
14,000	154	146	52 1/2	94	60 1/2	24	5	17 1/2	14 1/2	163,000
15,000	157	149 1/4	53 5/8	96	61 7/8	24	5	17 1/2	14 7/8	170,000
16,000	161	152 1/2	54 7/8	98 1/4	63 1/4	24	5	17 1/2	15 1/8	176,000
17,000	164	155 1/2	56	100	64 1/2	24	5	17 1/2	15 1/2	183,000
18,000	167	158 1/2	57	102	65 5/8	24	5	17 1/2	15 3/4	189,000
19,000	170 1/8	161 3/8	58	104	66 3/4	25	5 1/2	20	16	195,500
20,000	173	164	59	106	68	25	5 1/2	20	16 3/8	220,000
25,000	186 1/2	206	63 1/2	114	73 1/4	27	5 1/2	21 1/2	17 1/2	249,000
30,000	198	220	67 1/2	121 1/2	78	28 1/2	5 1/2	23	19 3/4	274,000
35,000	206 1/2	230	71	127 3/4	82	30	5 1/2	24	19 3/4	302,000
40,000	218	240	74 1/2	130 3/4	85 1/2	31 1/2	6	25	20 1/2	330,000

Figure 3-9. US Navy LWT anchor and specifications.

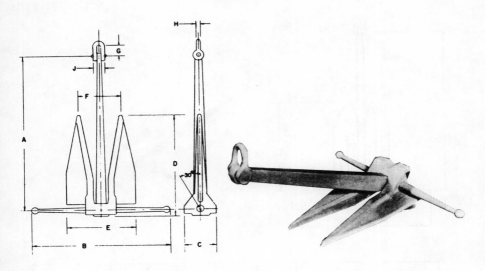

ANCHOR WEIGHT	A	B	C	D	E	F	G	H	J	PROOF TEST LBS
200	52 1/2	45 1/2	9 3/4	34 7/32	20 9/16	6 1/8	2 7/8	7/8	2 13/16	5,200
300	56	48 1/2	11	36 3/4	23 3/16	13 1/2	4 3/16	1 1/8	3 3/4	7,350
400	58	50	11 1/2	38 1/8	23 3/16	13 7/8	4 7/8	1 1/8	3 1/2	9,300
500	63	54 1/2	12 3/4	43	26 1/8	15 1/8	5 7/8	1 1/2	4 3/4	11,250
750	67 1/2	58 1/2	14 1/2	44 11/16	28 1/4	16 1/2	5 7/8	1 1/2	5	15,500
1,000	72	62 1/4	16 1/4	47 3/4	30	18 11/16	7 7/8	2	6 1/4	19,000
1,500	77	70 7/8	18 3/4	51 3/16	32	19 3/8	7 7/8	2	6 3/4	26,400
2,000	83	71 3/4	20 5/8	55 1/2	35 3/4	22 5/8	11 5/8	2 1/2	8 1/4	32,500
2,500	89	84 1/2	22	59	36 1/2	13 7/8	11 5/8	2 1/2	7 5/8	38,000
3,000	94	89 1/4	23 1/2	62 1/2	38 7/8	14 3/8	11 3/4	3	8 1/2	45,000
4,000	104	98 3/4	26	69 3/16	41 1/4	15 3/8	11 3/4	3	8 3/4	57,500
5,000	112	106 1/2	28	74 7/16	45 3/16	17 15/16	13 3/4	3 3/8	9 3/4	70,000
6,000	118	112 1/4	30	78 1/2	47 1/2	18 5/8	15 1/8	3 3/4	10 3/4	81,500
7,000	123	116 7/8	31 1/4	82 1/32	49 3/4	19 1/4	14 1/8	4	11 1/2	92,375
8,000	129	122 1/2	32 1/2	86 3/16	52 1/16	20 3/16	16 3/4	4 1/8	11 15/16	103,250
9,000	133	126 1/4	34	89	53 15/16	21 1/16	17 7/16	4 5/16	12 5/8	114,125
10,000	138	131	35	92 1/2	55 7/8	21 7/8	18 1/8	4 1/2	13 1/4	125,000
11,000	141 3/8	134	35 7/8	94 5/8	57 1/4	22 3/8	18 9/16	4 5/8	13 9/16	135,665
12,000	145	145	36 3/4	97 1/8	58 3/4	22 15/16	19	4 3/4	13 15/16	147,335
13,000	148 1/2	148 1/2	37 5/8	99 7/16	60 1/4	23 1/2	19 7/16	4 7/8	14 5/16	157,000
14,000	152	152	38 1/4	102	61 5/8	24	19 3/4	5	15 1/16	163,000
15,000	155 3/4	155 3/4	39 1/4	104 5/16	63 1/8	24 5/8	20 1/4	5 1/8	15 3/4	170,000
16,000	160	160	39 3/4	107 1/2	64 7/8	25 1/4	20 1/2	5 1/4	15 3/4	176,000
17,000	163	163	40 1/2	110 3/8	66 1/4	25 7/8	21 1/16	5 3/8	16 1/8	183,000
18,000	166	166	41 1/2	110 3/8	66 1/4	25 7.8	21 1/16	5 3/8	16 1/8	189,000
19,000	169	169	42	114 13/16	68 1/2	26 5/8	21 3/4	5 17/32	16 7/16	195,500
20,000	171 13/16	163 1/8	43 5/8	115 1/8	69 9/16	27 1/4	22 9/16	5 5/8	16 9/16	220,000
25,000	185	176	47	124	75	29 1/2	24 1/4	6	17 7/8	249,000
30,000	197	187	50	132	79 1/2	31 1/4	25 3/4	6 1/2	19	274,000

Figure 3-10. Danforth anchor and specifications.

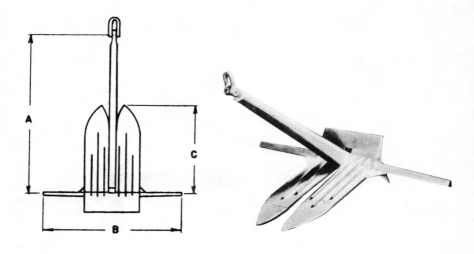

ANCHOR WEIGHT	A	B	C	SHACKLE SIZE	DESIGN H.P.	PROOF TEST LBS
200	42	59	26	3/4	6,000	9,000
3,000	129	109	72	1 3/4	60,000	90,000
6,000	144	143	92	2 1/4	120,000	180,000
9,000	160	170	100	2 3/4	180,000	270,000
12,000	186	197	113	3	210,000	315,000
15,000	204	224	126	3 1/2	240,000	360,000

*Proof test applied to shank only.

Figure 3-11. STATO anchor and specifications.

procedure tends to make the anchor dive (see Figure 3-12). To save time, the anchors can be "pre-set," that is, set before the drilling vessel arrives on location.

If the anchors will not hold a pretension determined by mooring calculations, tandem or "piggyback" anchors can be used. This is done by attaching the pendant line to the anchor shackle of another anchor and deploying it in a manner similar to the original anchor. Anchors are marked for recovery by a buoy attached to the pendant line.

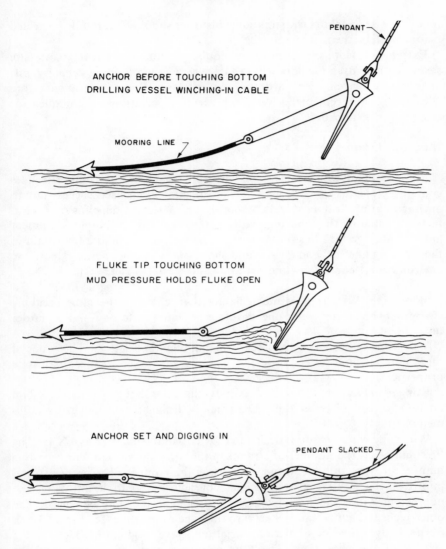

ANCHOR BEFORE TOUCHING BOTTOM
DRILLING VESSEL WINCHING-IN CABLE

PENDANT

MOORING LINE

FLUKE TIP TOUCHING BOTTOM
MUD PRESSURE HOLDS FLUKE OPEN

ANCHOR SET AND DIGGING IN

PENDANT SLACKED

Figure 3-12. The sequence of setting an anchor with a workboat.

It has been reported[9] that at times, tensions above 300 kips have been required to hold a vessel under certain environmental conditions. Because of this high tension, required reliability and mechanical punishment inherent in handling anchors, it is suggested that anchors be selected with a proof test that is at least 50% greater than the maximum anticipated holding capacity required. Used anchors with unknown histories should be carefully inspected before putting them into service.

Extreme sea floor conditions may require pile anchors. On very soft, silty seabeds where drag anchors will not develop enough holding capacity, pile anchors are driven. For hard rock or coral sea floors, pile anchors are "drilled-in" and cemented. Reese[10] presents calculational techniques for designing pile anchors.

Mooring Lines

Mooring lines which consist of wire rope or chain, or composite lines containing lengths of both wire and chain, have been used successfully to anchor floating drilling vessels. Mooring line selection depends on several factors, including the mooring loads, water depth, line handling equipment and line storage facilities on the vessel. The size, strength, and the length of the mooring lines depend on size and shape of the vessel, on water depth, on environmental conditions and on allowable vessel offset.

Wire rope. Wire rope is much lighter than chain of the same breaking strength. Therefore, wire rope develops a flatter catenary and a harder mooring system.

Wire rope is described by a series of numbers and letters. The order of these numbers and letters varies, so care must be taken to specify all rope properties.

Rope *diameter*, or *size*, is the maximum diameter that can be measured on an undeformed piece of rope. Diameter is designated in inches or millimeters.

Wire rope *construction* is designated by two numbers, such as 6 X 19. The first number is the number of strands in the rope; the second number is the number of wires in each strand. Increasing the number of wires per strand improves the flexibility and fatigue characteristics. In addition to the number of wires, the size of the wires in each strand may be varied. The objective is to pack as much steel into the rope as possible (see Figure 3-13). Surprisingly, these construction variations have negligible effect on the density of the wire rope. The four common types of construction are Warrington, Seale, filler wire and Warrington-Seale.

Classification of a wire rope is the grouping of wire ropes with similar construction and commonly-listed strengths and weights. The numbers, such as 6 X 19, are used for *classification* just as for *construction*. This is confusing if the grouping is not understood. For mooring a floating drilling vessel, two classifications are frequently used. The 6 X 19 classification ropes have 6 strands, with from 15 to 26 wires per strand (including filler wires) and not more than 12 outer wires per strand. Physical properties of the

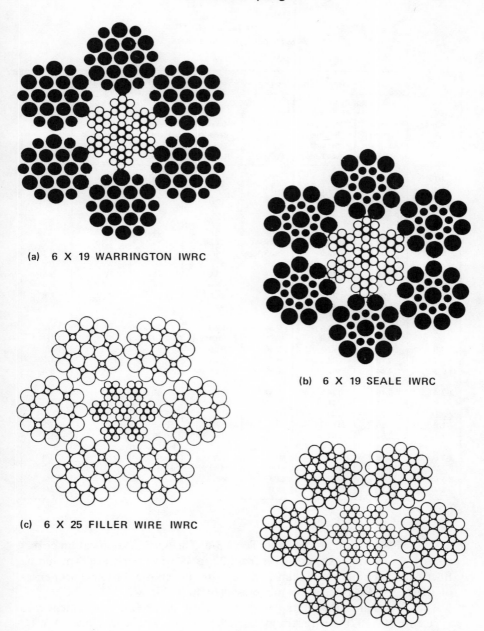

(a) 6 X 19 WARRINGTON IWRC

(b) 6 X 19 SEALE IWRC

(c) 6 X 25 FILLER WIRE IWRC

(d) 6 X 36 WARRINGTON–SEALE IWRC

Figure 3-13. Strand construction for mooring lines.

Table 3-6
Wire Rope Specifications
6 X 19 Bright

| Diameter Inches | IMPROVED PLOW STEEL | | | | EXTRA IMPROVED PLOW STEEL | | Diameter Inches |
| | FIBER CORE | | IWRC | | IWRC | | |
	Approx. Weight per Foot Pounds	Minimum Breaking Strength Net Tons	Approx. Weight per Foot Pounds	Minimum Breaking Strength Net Tons	Approx. Weight per Foot Pounds	Minimum Breaking Strength Net Tons	
3/16	.059	1.55	.065	1.67	—	—	3/16
1/4	.105	2.74	.116	2.94	.116	3.40	1/4
5/16	.164	4.26	.180	4.58	.180	5.27	5/16
3/8	.236	6.10	.260	6.56	.260	7.55	3/8
7/16	.32	8.27	.35	8.89	.35	10.2	7/16
1/2	.42	10.7	.46	11.50	.46	13.3	1/2
9/16	.53	13.5	.59	14.5	.59	16.8	9/16
5/8	.66	16.7	.72	17.9	.72	20.6	5/8
3/4	.95	23.8	1.04	25.6	1.04	29.4	3/4
7/8	1.29	32.2	1.42	34.6	1.42	39.8	7/8
1	1.68	41.8	1.85	44.9	1.85	51.7	1
1-1/8	2.13	52.6	2.34	56.5	2.34	64.0	1-1/8
1-1/4	2.63	64.6	2.89	69.4	2.89	79.9	1-1/4
1-3/8	3.18	77.7	3.50	83.5	3.50	96.	1-3/8
1-1/2	3.78	92.	4.16	98.9	4.16	114.	1-1/2
1-5/8	4.44	107.	4.88	115.	4.88	132.	1-5/8
1-3/4	5.15	124.	5.67	133.	5.67	153.	1-3/4
1-7/8	5.91	141.	6.50	152.	6.50	174.	1-7/8
2	6.72	160.	7.39	172.	7.39	198.	2
2-1/8	7.59	179.	8.35	192.	8.35	221.	2-1/8
2-1/4	8.51	200.	9.36	215.	9.36	247.	2-1/4
2-3/8	9.48	222.	10.4	239.	10.4	274.	2-3/8
2-1/2	10.5	244.	11.6	262.	11.6	302.	2-1/2
2-5/8	11.6	268.	12.8	288.	12.8	331.	2-5/8
2-3/4	12.7	292.	14.0	314.	14.0	361.	2-3/4

6 X 19 classification are listed in Table 3-6. The 6 x 37 classification covers wire ropes having 6 strands with from 27 to 49 wires per strand (including filler wires) and not more than 18 outer wires per strand. Physical properties of the 6 X 37 classification are presented in Table 3-7.

The *core*, that is, the internal part of the rope, may be an independent wire rope core (IWRC), a wire strand core (WSC), or a fiber core (fiber). IWRC is preferred for drilling vessels, because it is stronger and more fatigue resistant than the others.

Steel may be plow steel (PS), improved plow steel (IPS), or extra improved plow steel (EIPS). Either IPS or EIPS is recommended for mooring floaters.

Table 3-7
Wire Rope Specifications
6 X 37 Bright

| Diameter Inches | IMPROVED PLOW STEEL | | | | EXTRA IMPROVED PLOW STEEL | | Diameter Inches |
| | FIBER CORE | | IWRC | | IWRC | | |
	Approx. Weight per Foot Pounds	Minimum Breaking Strength Net Tons	Approx. Weight per Foot Pounds	Minimum Breaking Strength Net Tons	Approx. Weight per Foot Pounds	Minimum Breaking Strength Net Tons	
3/16	.059	1.46	—	—	—	—	3/16
1/4	.105	2.59	.116	2.78	.116	3.20	1/4
5/16	.164	4.03	.180	4.33	.180	4.98	5/16
3/8	.236	5.77	.260	6.20	.260	7.14	3/8
7/16	.32	7.82	.35	8.41	.35	9.67	7/16
1/2	.42	10.2	.46	11.0	.46	12.6	1/2
9/16	.53	12.9	.59	13.9	.59	15.9	9/16
5/8	.66	15.8	.72	17.0	.72	19.6	5/8
3/4	.95	22.6	1.04	24.3	1.04	27.9	3/4
7/8	1.29	30.6	1.42	32.9	1.42	37.8	7/8
1	1.68	39.8	1.85	42.8	1.85	49.1	1
1-1/8	2.13	50.1	2.34	53.9	2.34	61.9	1-1/8
1-1/4	2.63	61.5	2.89	66.1	2.89	76.1	1-1/4
1-3/8	3.18	74.1	3.50	79.7	3.50	91.7	1-3/8
1-1/2	3.78	87.9	4.16	94.5	4.16	108.	1-1/2
1-5/8	4.44	103.	4.88	111.	4.88	127.	1-5/8
1-3/4	5.15	119.	5.67	128.	5.67	146.	1-3/4
1-7/8	5.91	136.	6.50	146.	6.50	168.	1-7/8
2	6.72	154.	7.39	165.	7.39	190.	2
2-1/8	7.59	173.	8.35	186.	8.35	214.	2-1/8
2-1/4	8.51	193.	9.36	207.	9.36	239.	2-1/4
2-3/8	9.48	214.	10.4	230.	10.4	264.	2-3/8
2-1/2	10.5	236.	11.6	254.	11.6	292.	2-1/2
2-5/8	11.6	260.	12.8	279.	12.8	321.	2-5/8
2-3/4	12.7	284.	14.0	305.	14.0	350.	2-3/4
2-7/8	—	—	15.3	333.	15.3	382.	2-7/8
3	15.1	335.	16.6	360.	16.6	414.	3
3-1/8	16.4	362.	18.0	389.	18.0	448.	3-1/8
3-1/4	17.7	390.	19.5	419.	19.5	483.	3-1/4
3-3/8	—	—	21.0	451.	21.0	518.	3-3/8
3-1/2	20.6	449.	22.7	483.	22.7	555.	3-1/2

Steel *finish* is designated as *bright* for a plain steel finish. For mooring, the rope should be *galvanized* to improve its resistance to corrosion. Galvanizing reduces the strength of wire rope by about 10%.

Lay is the way the wires are twisted to form each strand, combined with how the strands are wound around the core. The *right regular lay* is recommended because it is less likely to kink and untwist than the *lang lay*, another common lay.

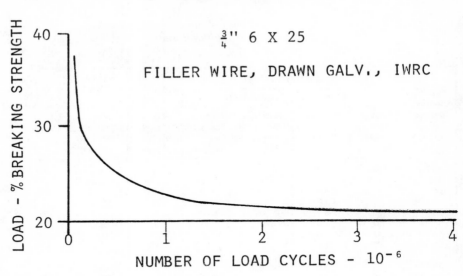

Figure 3-14. Effect of fatigue on 3/4-in. wire rope.

Preforming removes the tendency of the wires and strands to straighten out, and leaves them relaxed in their normal position in the rope. In preformed wire rope, the individual wires in the strand and the strands composing the rope are preshaped before they are assembled into the finished rope.

Care should be taken to specify the exact wire rope needed when ordering. The preceding means for designating the rope should be supplemented with the use for the rope, the length of each line and the line terminations.

For North Sea use, Arcaster[11] recommends 6 X 36/41/49 (37 Classification) IPS-IWRC, right regular lay, galvanizing class A (API). Warrington-Seale is a good construction for rough weather use, and preforming is always advisable on lines for mooring floating drilling vessels.

Wire rope fittings are usually subjected to the same loading as the rope, and should be equally as strong. Unfortunately, only swaged fittings are equal to rope strength, but they are not installed in the field. Zinc-poured sockets can be installed in the field, and when installed properly they will have about 95% of the catalogue breaking strength of the wire rope.

Wire rope is subjected to fatigue failure that is accelerated by the magnitude of the load. An increase in the load range can be expected to decrease the life of a mooring line. Figure 3-14 shows that a ¾-inch, 6 X 25 IWRC rope with a life of more than 4 X 10^6 cycles when loaded to 20% of the breaking strength has a life of 1 X 10^5 cycles when loaded to 30%. The useful

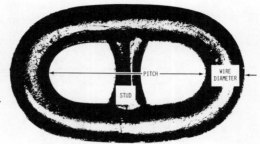

Figure 3-15. Chain nomenclature.

life decreases even more rapidly with increased loading above 30% of the breaking strength.

Wire rope fatigues more rapidly near the connector terminations than in the body of the rope. Consequently, it is advisable to cut off the last 15 feet of mooring line and reinstall the sockets at specific intervals. Such periodic maintenance will reduce the probability of fatigue failures in mooring lines.

Chain. Stud-link chain is used to moor drilling vessels. The chain *size* is designated by the *wire or bar diameter* (see Figure 3-15). The *pitch* is the inside length, and determines the fit of the chain in the wildcat. The *stud* is the supporting arch to keep the chain from collapsing.

Two types of mooring chain are in general use: flashwelded and dielocked chain. *Flashwelded* chain is manufactured in many parts of the world. *Dielocked* chain is a U.S. Navy development, and is manufactured only by Baldt Corporation, located in Chester, Pennsylvania. It is expensive and has a limited market.

Three grades of flashwelded chain—Grade 2, Grade 3 and "oil rig quality"—are commercially available. The higher grade indicates a higher strength. Weights and strengths of chains are shown in Table 3-8.

Specifications have been prepared by various testing societies for the graded chains. Societies specify the chemical composition of the material, the physical and material specifications, the dimensional tolerances, the minimum weight per link, and the proof and break test loading. *Oil rig quality* chain is not covered by any detailed specifications. Use of chain with this designation is not advisable, because the chain may have an initial high strength and proof strength but poor ductility and comparatively short fatigue life.

General dimensions and strength of oil rig quality, dielocked and flashwelded chain are shown in Table 3-8 and ABS testing requirements for grades 2 and 3 are shown in Table 3-9. The general dimensions for a chain

Table 3-8
Stud-link Chain, Dimensions and Data

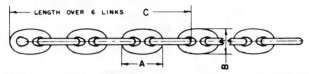

CHAIN SIZE IN INCHES				REGULAR DIE-LOCKED AND "OIL RIG" WELDED CHAIN				SUPER STRENGTH DIE-LOCKED CHAIN			
WIRE DIAM.	LENGTH OF LINK A	WIDTH OF LINK B	LENGTH OVER 6 LINKS C	WT. PER FT. OF CHAIN IN LBS. In Air	In Water	PROOF TEST, KIPS	BREAK TEST, KIPS	WT. PER FT. OF CHAIN IN LBS. In Air	In Water	PROOF TEST, KIPS	BREAK TEST, KIPS
1	6	3 9/16	26	10.0	8.7	84.0	129.0	12.5	10.8		157.3
1 1/16	6 3/8	3 3/4	27 5/8	11.3	9.8	95.0	145.0	14.1	12.3		176.9
1 1/8	6 3/4	4	29 1/4	12.7	11.0	106.0	161.0	15.9	13.8		196.4
1 3/16	7 1/8	4 1/4	30 7/8	14.2	12.3	118.0	179.5	17.8	15.5		218.9
1 1/4	7 1/2	4 1/2	32 1/2	15.7	13.6	130.0	198.0	19.6	17.1		241.5
1 5/16	7 7/8	4 3/4	34 1/4	17.3	15.1	143.5	216.5	21.6	18.8		264.1
1 3/8	8 1/4	4 15/16	35 3/4	19.0	16.5	157.0	235.0	23.8	20.7		286.7
1 7/16	8 5/8	5 3/16	37 3/8	20.7	18.0	171.0	257.5	25.9	22.5		314.1
1 1/2	9	5 3/8	39	22.6	19.6	185.0	280.0	28.2	24.5		341.6
1 9/16	9 3/8	5 5/8	40 5/8	24.4	21.2	200.5	302.5	30.5	26.5		369.0
1 5/8	9 3/4	5 7/8	42 1/4	26.1	22.7	216.0	325.0	32.6	28.4		396.5
1 11/16	10 1/8	6 1/16	43 7/8	28.1	24.4	232.5	352.5	35.1	30.5		430.0
1 3/4	10 1/2	6 5/16	45 1/2	30.0	26.1	249.0	380.0	37.5	32.6		463.6
1 13/16	10 7/8	6 1/2	47 1/8	32.6	28.3	267.0	406.0	40.8	35.5		495.3
1 7/8	11 1/4	6 3/4	48 3/4	34.8	30.3	285.0	432.0	43.5	37.8		527.0
1 15/16	11 5/8	7	50 3/8	37.0	32.2	303.5	460.5	46.2	40.2		561.2
2	12	7 3/16	52	39.2	34.1	322.0	488.0	49.0	42.6		595.3
2 1/16	12 3/8	7 7/16	53 5/8	41.7	36.3	342.0	518.0	52.1	45.3		631.9
2 1/8	12 3/4	7 5/8	55 1/4	44.2	38.4	362.0	548.0	55.3	48.1		668.6
2 3/16	13 1/8	7 7/8	56 7/8	46.9	40.8	382.5	579.1	58.5	50.8		706.5
2 1/4	13 1/2	8 1/8	58 1/2	49.6	43.1	403.0	610.0	62.0	53.9		744.2
2 5/16	13 7/8	8 5/16	60 1/8	52.4	45.5	425.0	642.5	65.5	57.0		783.8
2 3/8	14 1/4	8 9/16	61 3/4	55.1	47.9	447.0	675.0	68.9	59.9		823.5
2 7/16	14 5/8	8 3/4	63 3/8	58.0	50.4	469.5	709.5	72.5	63.0		865.5
2 1/2	15	9	65	61.5	53.5	492.0	744.0	76.8	66.8		907.6
2 9/16	15 3/8	9 1/4	66 5/8	64.5	56.0	516.0	778.5	80.6	70.1		949.7
2 5/8	15 3/4	9 7/16	68 1/4	67.9	59.0	540.0	813.0	84.9	73.9		991.8
2 11/16	16 1/8	9 11/16	69 7/8	71.2	62.0	565.0	849.0	89.0	77.4		1035.7
2 3/4	16 1/2	9 7/8	71 1/2	74.7	64.9	590.0	885.0	93.3	81.1		1079.7
2 13/16	16 7/8	10 1/8	73 1/8	78.2	68.0	615.0	925.0	97.7	85.0		1128.5
2 7/8	17 1/4	10 3/8	74 3/4	81.9	71.1	640.0	965.0	102.3	89.1		1177.3
2 15/16	17 5/8	10 9/16	76 3/8	85.6	74.4	666.5	1005.0	107.0	93.0		1226.1
3	18	10 13/16	78	89.3	77.6	693.0	1045.0	111.7	97.0		1274.9
3 1/16	18 3/8	11	79 5/8	93.0	80.9	720.5	1086.5	116.2	101.0		1324.9
3 1/8	18 3/4	11 1/4	81 1/4	97.0	84.3	748.0	1128.0	121.2	105.3		1376.1
3 3/16	19 1/8	11 1/2	82 7/8	101.0	88.0	776.0	1169.0	126.1	109.8		1426.1
3 1/4	19 1/2	11 11/16	84 1/2	105.1	91.4	804.0	1210.0	131.3	114.2		1476.2
3 5/16	19 7/8	11 15/16	86 1/8	109.1	95.0	833.0	1253.0	136.3	118.5		1528.6
3 3/8	20 1/4	12 1/8	87 3/4	113.5	98.6	862.2	1296.0	141.8	123.2		1581.1
3 7/16	20 5/8	12 3/8	89 3/8	116.5	101.3	892.0	1339.5	145.5	126.5		1634.2
3 1/2	21	12 5/8	91	122.2	106.2	922.0	1383.0	152.8	132.8		1687.3
3 5/8	21 3/4	12 15/16	94 1/4	129.0	112.1	1021.0	1566.0	161.2	140.2		1910.5
3 3/4	22 1/2	13 3/8	97 1/2	145.0	122.1	1120.0	1750.0	175.8	152.8		2135.0
3 7/8	23 1/4	14	100 3/4	148.5	129.0	1205.0	1836.4	185.5	161.0		2240.0
4	24	14 3/8	104	156.5	136.0	1298.0	1996.5	196.0	170.0		2435.0
4 1/8	24 3/4	14 7/8	107 1/4	166.8	145.0	1347.4	2062.5	208.3	181.5		2603.5
4 1/4	25 1/2	15 5/16	110 1/2	176.8	153.8	1393.7	2134.0	221.0	192.2		2764.5
4 3/8	26 1/4	15 3/4	113 3/4	187.4	163.0	1569.7	2398.0	234.0	203.5		2925.5
4 1/2	27	16 3/16	117	198.2	172.5	1672.0	2508.0	248.0	216.0		3059.8
4 5/8	27 3/4	16 5/8	120 1/4	209.4	182.0	1775.0	2675.0	261.6	227.5		3260.0
4 3/4	28 1/2	17 1/8	123 1/2	220.4	192.0	1870.0	2805.0	276.0	240.0		3420.0

(Super Strength Proof Test column labeled vertically: REGULAR DIE-LOCKED CHAIN / SAME AS REGULAR DIE-LOCKED CHAIN)

link are multiples of the nominal wire diameter. The pitch is equal to 4 wire diameters, the outside length is 6 wire diameters, the gauge test length is six links of chain or 26 wire diameters. The gauge length of a chain is very important in determining if the chain will fit a standard wildcat.

Table 3-9
American Bureau of Shipping
Testing Requirements

Diameter	Grade 2 Proof Load	Grade 2 Breaking Load	Grade 3 Proof Load	Grade 3 Breaking Load	Minimum weight pounds 15 Fathoms	Diameter	Grade 2 Proof Load	Grade 2 Breaking Load	Grade 3 Proof Load	Grade 3 Breaking Load	Minimum weight pounds 15 Fathoms
inches	pounds	pounds	pounds	pounds		inches	pounds	pounds	pounds	pounds	
1/2	15275	21390	21390	30555	225	2 3/8	314300	439900	439900	628400	4725
9/16	19285	26995	26995	38575	290	2 7/16	330000	461900	461900	659800	4960
5/8	23745	33220	33220	47465	365	2 1/2	346000	484300	484300	691800	5265
11/16	28625	40095	40095	57275	405	2 9/16	363300	507200	507200	725600	5535
3/4	33980	47580	47580	67960	480	2 5/8	378900	530400	530400	757800	5815
13/16	39760	55665	55665	79520	570	2 11/16	395800	554200	554200	791600	6105
7/8	45965	64355	64355	91840	655	2 3/4	413100	578400	578400	826200	6405
15/16	52620	73650	73650	105210	755	2 13/16	430700	603000	603000	861400	6705
1	59695	83550	83550	119390	855	2 7/8	448600	628000	628000	897000	7015
1 1/16	67155	94055	94055	134850	970	2 15/16	466700	653500	653500	933500	7330
1 1/8	75040	105050	105050	150080	1085	3	485200	679200	679200	970300	7650
1 3/16	83440	116700	116700	166770	1215	3 1/16	503900	705400	705400	1007700	7980
1 1/4	92180	129020	129020	184240	1345	3 1/8	522900	732100	732100	1045800	8320
1 5/16	101250	141800	141800	202500	1485	3 3/16	542200	759000	759000	1084300	8660
1 3/8	110810	155200	155200	221600	1625	3 1/4	561700	786500	786500	1123500	9010
1 7/16	120740	169000	169000	241470	1775	3 5/16	581500	814100	814100	1163100	9360
1 1/2	131040	183450	183450	262080	1935	3 3/8	601700	842500	842500	1203500	9725
1 9/16	141800	198500	198500	283600	2090	3 7/16	622000	870800	870800	1244000	10095
1 5/8	152880	213920	213920	305650	2235	3 1/2	642700	899900	899900	1285400	10475
1 11/16	166660	229090	229090	326700	2410	3 9/16	663500	928800	928800	1326900	10860
1 3/4	176180	246620	246620	352240	2590	3 5/8	684500	958400	958400	1369200	11250
1 13/16	188380	263600	263600	376600	2785	3 3/4	727600	1018500	1018500	1455000	12025
1 7/8	200900	281300	281300	401900	2975	3 7/8	771500	1080000	1080000	1542800	12850
1 15/16	213600	299300	299300	426900	3175	3 15/16	793700	1111100	1111100	1587400	13275
2	227100	318100	318100	454400	3355	4	816100	1142700	1142700	1632400	13700
2 1/16	240800	337100	337100	481600	3570	4 1/8	861800	1206600	1206600	1723700	14560
2 1/8	254800	356700	356700	509600	3785	4 1/4	908300	1271800	1271800	1816800	15350
2 3/16	269100	376900	376900	538300	4015	4 3/8	955700	1338000	1338000	1911300	16200
2 1/4	283900	396400	396400	569700	4245	4 1/2	1003700	1405300	1405300	2007500	17100
2 5/16	298900	418400	418400	597700	4485	4 5/8	1052600	1473600	1473600	2105200	18000
						4 3/4	1102100	1542900	1542900	2204200	18900

Chain will fatigue and fail as a result of cyclic loading. The number of cycles to failure depends on the loading. Higher average tensions and greater stress ranges result in a lower chain life. A rule of thumb with metals is to keep the tension below $1/3$ of the yield point to prolong the fatigue life. A paper published by Childers[12] indicates that this rule may be valid for chain. He stated that the higher the load (especially over one-third the breaking strength), and the larger the spread between low and high tensions during cycling, the shorter the life.

Table 3-10
Table for Renewing Stud-Link Chain

Original Diameter (in.)	Renewal Diameter (in.)	Original Diameter (in.)	Renewal Diameter (in.)
1 3/4	1 9/16	2 11/16	2 13/32
1 13/16	1 5/8	2 3/4	2 15/32
1 7/8	1 11/16	2 13/16	2 17/32
1 15/16	1 23/32	2 7/8	2 9/16
2	1 25/32	2 15/16	2 5/8
2 1/16	1 27/32	3	2 11/16
2 1/8	1 29/32	3 1/16	2 3/4
2 3/16	1 31/32	3 1/8	2 13/16
2 1/4	2	3 3/16	2 7/8
2 5/16	2 1/16	3 1/4	2 15/16
2 3/8	2 1/8	3 5/16	2 31/32
2 7/16	2 3/16	3 3/8	3 1/32
2 1/2	2 1/4	3 7/16	3 1/16
2 9/16	2 9/32	3 1/2	3 1/8
2 5/8	2 11/32		

Fortunately, many chain defects can be detected visually prior to failure. This requires periodic inspections. The common area for fatigue cracks is 45° on either side of the link center line. All links with cracks should be cut out of the system. The stud in flashwelded links is held in place by friction. If the link has been deformed, the stud may become loose and even fall out. When inspecting a chain, the studs should be checked for tightness by tapping them with a hammer. Chain with loose studs can be repaired by welding the stud to the side of the link opposite the flashweld. Only the one side should be welded. The chain link should also be checked to determine if the chain has been excessively deformed or corroded. Table 3-10 shows the recommended renewal diameters for undersized links.

Although the proof strength of chain connecting links is greater than the same size chain link, all commercial connectors have a lower fatigue life than the chain. Consequently, they should be replaced more often than the chain. Swivels are commercially available but are highly susceptible to fatigue failure and should not be used for mooring a drilling vessel.

Chain is attached to the anchor by an end shackle (Figure 3-16) and an anchor shackle (Figure 3-17). If the mooring is a composite chain and wire rope system, it may be connected as in Figure 3-18.

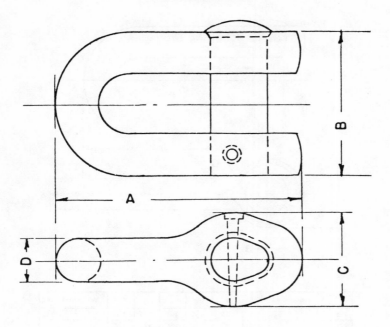

SIZE CHAIN	A	B	C	D	PROOF TEST IN POUNDS	BREAK TEST IN POUNDS
$\frac{1}{2}$ — $\frac{5}{8}$	$5\frac{3}{16}$	$3\frac{1}{4}$	2	$\frac{13}{16}$	32,300	52,200
$\frac{11}{16}$ — $\frac{3}{4}$	$6\frac{1}{4}$	4	$2\frac{3}{8}$	1	48,000	75,000
$\frac{13}{16}$ — $\frac{7}{8}$	$7\frac{1}{4}$	$4\frac{1}{2}$	$2\frac{5}{8}$	$1\frac{1}{8}$	64,000	98,000
$\frac{15}{16}$ — 1	$7\frac{7}{8}$	$4\frac{7}{8}$	3	$1\frac{1}{4}$	84,000	129,000
$1\frac{1}{16}$ — $1\frac{1}{8}$	$8\frac{5}{8}$	$5\frac{3}{8}$	$3\frac{3}{8}$	$1\frac{3}{8}$	106,000	161,000
$1\frac{3}{16}$ — $1\frac{1}{4}$	$9\frac{3}{8}$	$5\frac{7}{8}$	$3\frac{5}{8}$	$1\frac{1}{2}$	130,000	198,000
$1\frac{5}{16}$ — $1\frac{3}{8}$	$10\frac{3}{4}$	$6\frac{1}{2}$	$4\frac{1}{4}$	$1\frac{3}{4}$	157,000	235,000
$1\frac{7}{16}$ — $1\frac{1}{2}$	$11\frac{1}{2}$	$6\frac{7}{8}$	$4\frac{5}{8}$	$1\frac{7}{8}$	185,000	280,000
$1\frac{9}{16}$ — $1\frac{5}{8}$	$12\frac{1}{4}$	$7\frac{3}{8}$	5	2	216,000	325,000
$1\frac{11}{16}$ — $1\frac{3}{4}$	$13\frac{1}{8}$	$8\frac{1}{8}$	$5\frac{1}{4}$	$2\frac{1}{4}$	249,000	380,000
$1\frac{13}{16}$ — $1\frac{7}{8}$	$14\frac{5}{8}$	$8\frac{3}{4}$	6	$2\frac{1}{2}$	285,000	432,000
$1\frac{15}{16}$ — $2\frac{1}{8}$	$15\frac{7}{8}$	$9\frac{1}{2}$	$6\frac{5}{8}$	$2\frac{3}{4}$	362,000	548,000
$2\frac{3}{16}$ — $2\frac{3}{8}$	$17\frac{3}{8}$	$10\frac{3}{8}$	$7\frac{1}{4}$	3	447,000	675,000
$2\frac{7}{16}$ — $2\frac{5}{8}$	$19\frac{1}{8}$	$11\frac{1}{8}$	8	$3\frac{1}{4}$	540,000	813,000
$2\frac{11}{16}$ — $2\frac{7}{8}$	$20\frac{1}{2}$	$12\frac{3}{8}$	$8\frac{5}{8}$	$3\frac{1}{2}$	640,000	965,000
3 — $3\frac{1}{8}$	$22\frac{5}{8}$	$13\frac{1}{4}$	$9\frac{5}{8}$	4	748,000	1,128,000
$3\frac{1}{4}$ — $3\frac{3}{8}$	24	$13\frac{7}{8}$	$10\frac{1}{4}$	$4\frac{1}{4}$	862,200	1,296,000
$3\frac{1}{2}$	$25\frac{1}{2}$	$14\frac{3}{4}$	$10\frac{7}{8}$	$4\frac{1}{2}$	922,000	1,383,100
$3\frac{3}{4}$	$27\frac{7}{8}$	16	$11\frac{3}{4}$	$4\frac{7}{8}$	1,120,000	1,750,000

Figure 3-16. End shackle and typical data.

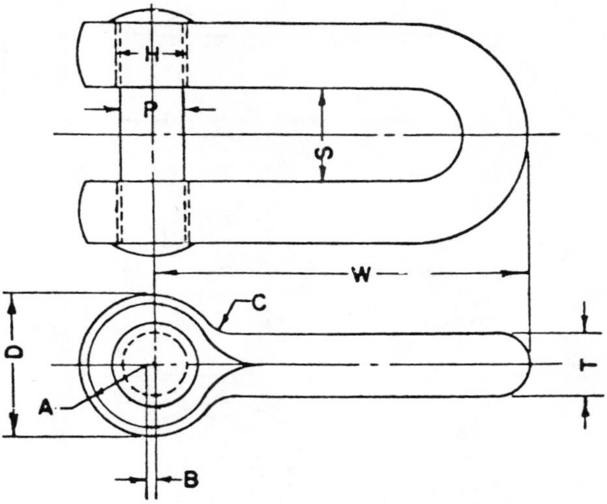

SHACKLE SIZE	A	B	C	D	H	P	S	T -	W	WT. LBS.
1	1 1/4	1/8	1	2 1/2	1 1/8	1	2 1/4	1	6 1/2	5 3/4
1 1/4	1 11/16	1/8	1 1/4	3 3/8	1 5/8	1 1/2	2 3/4	1 1/4	8 1/4	11
1 1/2	1 13/16	1/8	1 1/2	3 5/8	1 5/8	1 1/2	2 3/4	1 1/2	9	16 1/4
1 3/4	2 1/4	3/16	1 3/4	4 1/2	2 1/4	2	3 1/4	1 3/4	10 1/4	29
2	2 7/16	1/4	2	4 7/8	2 1/4	2	3 3/4	2	11 3/4	43
2 1/4	2 13/16	5/16	2 1/4	5 5/8	2 3/4	2 1/2	4 1/4	2 1/4	13 1/4	59
2 1/2	3	5/16	2 1/2	6	2 3/4	2 1/2	4 1/4	2 1/2	14 1/2	78
3	3 5/8	3/8	3	7 1/4	3 1/4	3	4 3/4	3	18	136
3 1/2	3 29/32	7/16	3 1/2	7 13/16	3 3/4	3 1/2	5 1/4	3 1/2	20 1/2	207
4	4 3/4	7/16	4	9 1/2	4 1/4	4	5 3/4	4	23	339
4 1/2	5 3/8	1/2	4 1/2	10 3/4	4 3/4	4 1/2	6 3/4	4 1/2	26 1/2	466
5	5 27/32	1/2	5	11 11/16	5 1/4	5	7 1/4	5	29	625

Figure 3-17. U-type anchor shackle and typical data.

Deck Machinery

Wire rope mooring lines require a winch and outboard fairleads. An arrangement is shown in Figure 3-19. Drum capacity and minimum distance to the fairlead sheave is shown in Figure 3-20. The fleet angle as defined in the figure should not be greater than 1½° for a smooth drum or 2° for a grooved drum. The radius of the rollers of sheaves in wire line fairleads should not be less than about 200 times the diameter of the outer wires of a strand.

A common wire rope tensiometer similar to the drilling line weight gauge is used. Tension can be read when the winch is moving slowly.

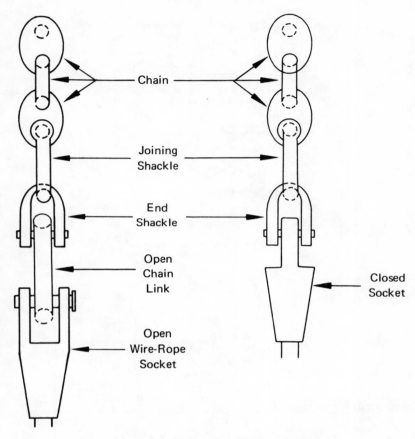

Figure 3-18. Typical wire rope connection to chain.

Chain mooring requires a wildcat, chain stopper, and fairleads, as shown in Figures 3-21 and 3-22. Chain tension is typically measured by a load cell or similar device in the bow stopper. However, this requires that the chain be stopped and braked before the tension can be measured. Measuring the tension while taking in or paying out the chain requires more sophisticated techniques than those commercially available.

Dynamic Positioning

A dynamically positioned vessel uses thrusters combined with propulsion screws to hold station instead of using a conventional mooring system. Dynamic positioning (DP) was employed on coring vessels in the 1950s, and

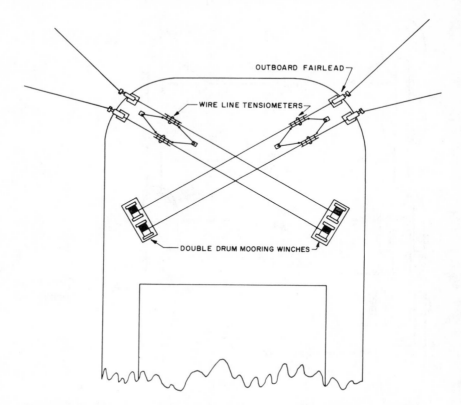

Figure 3-19. Example of wire rope deck machinery location for a barge.

a deepwater coring vessel, the Glomar Challanger, was commissioned in 1968. For commercial drilling, the first DP vessel, the SEDCO 445,[13] began shakedown in 1971 and was followed about six months later by the Pelican.[14] These vessels, along with other DP vessels continuously joining the fleet, are demonstrating their capability to hold station for two to four months while a well is being drilled. The SEDCO 445 set a record for dynamically positioned vessels of 2,298 feet while drilling off the African coast.[15] This deep water record has been exceeded by the Discoverer 534 (commissioned in 1975) with wells drilled in up to 3,461 feet of water while moored.[16] The Discoverer 534 has both mooring and DP capabilities. It is evident that the frontiers of deepwater drilling are being extended by both moored and dynamically positioned vessels.

The major advantage of dynamic positioning at present is mobility: no anchors to set or retrieve. Dynamically positioned vessels are particularly

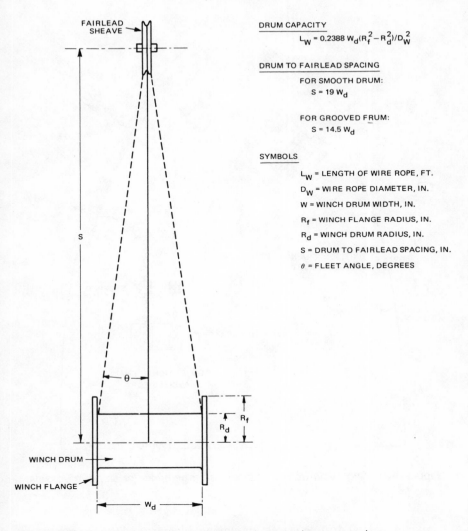

FAIRLEAD
SHEAVE

DRUM CAPACITY

$$L_W = 0.2388 \, W_d(R_f^2 - R_d^2)/D_W^2$$

DRUM TO FAIRLEAD SPACING

FOR SMOOTH DRUM:

$S = 19 \, W_d$

FOR GROOVED FRUM:

$S = 14.5 \, W_d$

SYMBOLS

L_W = LENGTH OF WIRE ROPE, FT.

D_W = WIRE ROPE DIAMETER, IN.

W = WINCH DRUM WIDTH, IN.

R_f = WINCH FLANGE RADIUS, IN.

R_d = WINCH DRUM RADIUS, IN.

S = DRUM TO FAIRLEAD SPACING, IN.

θ = FLEET ANGLE, DEGREES

S

θ

R_f

R_d

WINCH DRUM

WINCH FLANGE

W_d

Figure 3-20. Drum capacity and minimum drum-to-sheave spacing.

popular off the coast of Labrador where occasional iceberg dodging is required.[17] Eventually, DP will extend water depth capability beyond the realm of conventional mooring. The Glomar Challanger has cored in over 22,000 feet of water.[18] Commercial drilling, however, requires a riser system and ocean floor preventers that also limit water depth under current technology.

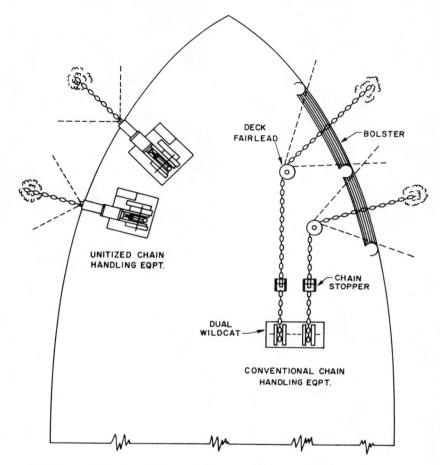

Figure 3-21. Two chain deck machinery arrangements for ship-like vessels.

All commercial drilling vessels employing dynamic positioning use acoustics as the primary source for determining the position of the vessel relative to the wellhead. This is the heart of any DP system and will be discussed next.

Position Referencing

The position of the rotary table (kelly bushing) relative to the center line of the well bore is determined by acoustic position referencing. The acoustic systems use at least three hydrophones mounted on the drilling vessel to triangulate on one or more acoustic beacons located on or near the wellhead.

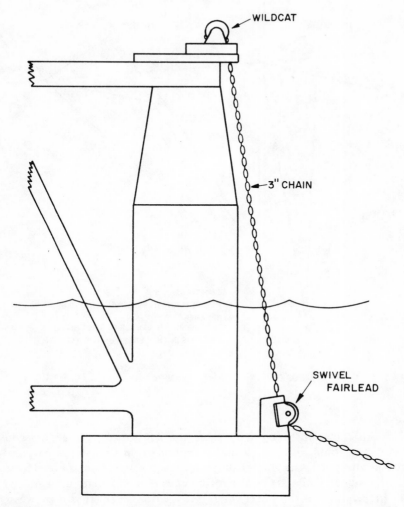

Figure 3-22. Typical chain wildcat and fairlead locations on a semi.

To understand the operating principles of acoustic position referencing, assume that:

1. The vessel is an equilateral triangle.
2. The kelly bushing (KB) is in the geometric center of the vessel.
3. The hydrophones are located at the points of the triangular vessel, that is, equidistant from the KB.
4. The subsea beacon is in the center of the well.
5. No pitch, no roll, no yaw and no heave are permitted.

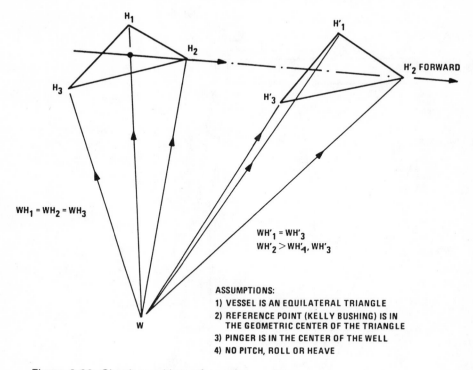

H_1

H_2

H'_1

H'_2 FORWARD

H_3

H'_3

$WH_1 = WH_2 = WH_3$

$WH'_1 = WH'_3$
$WH'_2 > WH'_4, WH'_3$

W

ASSUMPTIONS:
1) VESSEL IS AN EQUILATERAL TRIANGLE
2) REFERENCE POINT (KELLY BUSHING) IS IN
   THE GEOMETRIC CENTER OF THE TRIANGLE
3) PINGER IS IN THE CENTER OF THE WELL
4) NO PITCH, ROLL OR HEAVE

Figure 3-23. Simple position-referencing system.

Now, as shown in Figure 3-23, when the KB is directly over the acoustic beacon, the sound of a single pulse emitted from the subsea beacon will arrive at all three hydrophones simultaneously. If the vessel is offset forward *only*, the signal would arrive at hydrophones H_1 and H_3 simultaneously and later (depending on the offset) at H_2. It is also evident that if the vessel is offset forward and starboard, then hydrophones H_2 and H_3 will be farther away from the pinger than H_1 and the pulses will arrive at H_2 and H_3 later depending upon the magnitude of the offset. The position X any Y can be calculated by knowing:

1. The difference in times of arrival of a single pulse at the different hydrophones
2. The distance between these hydrophones
3. The speed of sound in water
4. The water depth

In fact, for the case with the beacon in the center of the well, yaw can be permitted, because all calculations would be in *ship*'s coordinates (fore/aft, port/starboard). Since beacons will not survive in the well bore and must be offset, the position reference system must be aware of *earth's coordinates* (north/south, east/west). This and other aspects of the ideal situation explained above are altered by information put into the device by hand or by external sensing equipment.

Corrections are required for pitch and roll, that is the inclination of the hydrophone plane. The inclination of the vessel is measured by a vertical reference unit consisting of a pendulum or gyroscope. This is often a weak link in the system, and some of the "electronic jitter" in various commercial systems are caused by the vertical reference unit.

Two types of subsea beacons are employed: a free-running pinger that continuously transmits signals at a specified rep rate, or a transponder that responds only to a command from the surface. The transponder is more accurate because it measures the water depth and requires no additional measurement for tides. In dynamic positioning, a transponder will allow the computer to select the most advantageous rep rate for the particular location.

The receiving hydrophones on the vessel must be carefully installed in dry dock with the vessel perfectly level. The tolerance allowed for this installation varies between manufacturers and recommendations by the manufacturer should be followed carefully.

In 1975, a unit incorporating three receiving hydrophones within a single casing was tested. This equipment has the advantage of being more portable and requiring less hull penetrations than its predecessors. The equipment passed a short field test; however, it is too early (as of 1979) for its reliability in extended field operations to be demonstrated to industry.

Backup position referencing systems are used in case of an interruption of the acoustic signal. The most popular system is taut wire. Taut wire systems are wires stretched from the vessel to a weight on the seabed. The inclination of this wire is measured to determine the angle of offset. Absolute values of taut wire positioning are not accurate, because subsea currents will deflect the wire. Thus, periodic calibration of the taut wire by comparing it with the acoustic positioning system is necessary. When properly calibrated, the taut wire is acceptable for dynamic positioning. Taut wire has been demonstrated to work in about 3,000 feet of water. For backup systems beyond the practical limits of taut wire, both inertial navigation and doppler sonar have been proposed. Both systems have a tendency to drift (i.e., have errors that grow with time) and, as with the taut wire, require periodic calibration with the acoustic system. As of 1979, neither inertial navigation nor doppler sonar had been demonstrated thoroughly to industry.

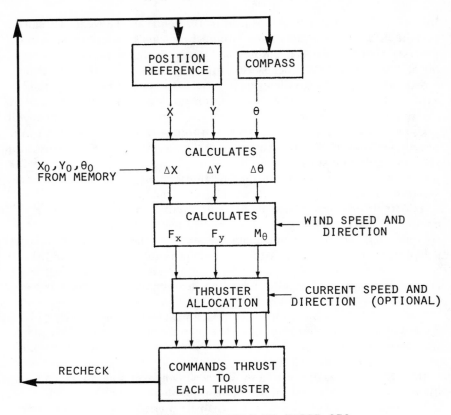

LAPSED TIME, ONE-HALF TO THREE SEC

Figure 3-24. Diagram of controller operations.

Controller Operations

The DP controller is a computer that takes input data from various sensors and estimates the thrust required to correct the vessel's position. A diagram of the logical sequence of events is shown in Figure 3-24 and is described below:

1. The computer receives the vessel position from the position referencing system, and the heading from the gyrocompass.
2. The computer compares the position and heading of the vessel with X_0 Y_0, and θ_0 that would place the vessel directly over the well and maintain the proper heading.

3. The forces required to bring the vessel over the well are estimated. Vessel position, heading and wind data are used, as will be explained.
4. The controller allocates thrust to the individual thrusters. To do this, the following procedure is used:

   a. Determine which thrusters are available.

   b. Calculate thruster efficiency if current data are available.

   c. Decide which thrusters to use for moment. Available thrusters farthest from the center of gravity will be used first.

   d. Determine lateral and forward thrust based on net thrust still required and power available.

5. Send the commands to the thrusters and check to see if they responded properly.
6. Go back and repeat the cycle.

Force estimations. Force estimations are based on two types of information, feedback and feedforward data.

Feedforward data are used to estimate the force required to resist wind gusts. Before the vessel is built for dynamic positioning, wind tunnel tests are run to determine the forces and moment that wind will impose on the vessel. Tables of wind effects at different angles and velocities are developed and stored in the computer memory. In operation, the wind speed and direction are measured, and the resulting forces and moment are estimated from the tables, so that the thruster forces can be adjusted for wind gusts. Steady-state wind forces are not the purpose of the wind information, because they are taken into consideration in the feedback equation.

Feedback information does not anticipate a force as the feedforward information, but merely reacts to changes in heading and position. For example, if a wave moves the vessel, the force is estimated based on the vessel's response to that wave and the other forces imposed on the rig. Now let us pick one of the forces F_x, F_y, *moment M_θ* and look at the feedback equation for estimating these forces and moment.

The offset in the X direction $= X - X_0 = \Delta X$, and,

force X direction	proportional term	differential term	integral term
F_x =	$A_x \Delta X$ +	$B_x \dfrac{d\Delta X}{dt}$ +	$C_x \int \Delta X \, dt$

A, B and C are dynamic response factors called *gains*, and can be considered analogous to the hardness of a mooring system. Higher gains will cause the system to be harder. Gains will vary with vessel mass and water depth. To give a feeling for the terms in the feedback equation, they are discussed below:

$A_x \Delta X$ is the proportional term, and increases the restoring thrust as the offset increases.

$B_x \dfrac{d\Delta X}{dt}$ is the differential term that increases the thrust relative to the speed at which the vessel is moving away from the well. This term may reverse the thrust as the well is approached, depending on the vessel speed.

$C_x \int \Delta X \; dt$ is the integral term. Without this term, the vessel would wander back and forth over the well and would not settle down. This term gains experience and determines the steady state forces required to hold the vessel over the well. Time is required to build up this integral when the system is first put into use. You can expect some wander when the vessel initially acquires position.

Thrusters

Two types of thrusters are used for dynamic positioning. *Constant pitch*, variable rpm units with DC motors, are the old standby of the shipping industry. They are very efficient and economical when the rpm is kept nearly constant, as on a cruise. In dynamic positioning, however, the thrust must be varied rapidly at times, which tends to decrease the efficiency. DC motors are limited in their ability to reverse thrust rapidly.

The second type of thruster is the *variable pitch*, constant rpm thruster. Variable pitch thrusters use the same principle as airplane propellers. They use AC motors, and the pitch is controlled hydraulically. These thrusters can reverse thrust more rapidly than can the DC drive motors, but some of the large variable pitch units have experienced hydraulic problems.

The variable pitch thrusters are very efficient for DP, and are used even for the main propulsion screws on the IHC-manufactured vessels.

Lateral thrusters have been made for up to 3,000-shaft HP with an efficiency on the order of 30 lb-thrust per HP. This efficiency will vary considerably with rpm, pitch and mounting on the vessel. Current velocities and direction relative to the axis of thrust appreciably effect the efficiency of a thruster. This is the reason that some controller manufacturers include current when allocating thrust to the individual units.

Obviously, a DP system is quite complicated. Many things require consideration, but the most important things are reliability and good experience of the manufacturers of the central DP controller, the thruster control mechanism, and the thrusters themselves. *Good experience means the demonstrated reliability of existing equipment and the availability of service for these systems.*

References

1. *Rules for Building and Classing Offshore Mobile Drilling Units*, American Bureau of Shipping. New York, 1973.

2. Bretschneider, C.L., "Water Loads on Fixed, Rigid Marine Structures," *Handbook of Ocean and Underwater Engineering*. McGraw-Hill, New York, N.Y., 1969, pp 12-24.

3. Kingsman, Blair, *Wind Waves*. Prentice-Hall, Englewood Cliffs, 1965.

4. Harris, L.M., *An Introduction to Deepwater Floating Drilling Operations*. Petroleum Publishing Co., Tulsa, 1972.

5. Gaskell, R.E., *Engineering Mathematics*. Henry Holt and Co., New York, 1958. pp 177-182.

6. Adams, R.B., "Analysis of Spread Moorings by Dimensionless Functions," OTC paper 1077. Presented at the Offshore Technology Conference, Houston, 1969.

7. Klotz, J.A., "Vessel Motion Control Through Anchor Line Springiness," OTC paper 1023. Presented at the Offshore Technology Conference, Houston, 1969.

8. Sisk, J.M., "An Analysis of Discoverer II Motions and Mooring Line Forces," OTC paper 1220. Presented at the Offshore Technology Conference, Houston, 1970.

9. Beck, R.W., "Anchor Performance Tests," OTC paper 1573. Presented at the Offshore Technology Conference, Houston, 1972.

10. Reese, L.C., "A Design Method for an Anchor Pile in a Mooring System," OTC paper 1745. Presented at the Offshore Technology Conference, Houston, 1973.

11. Ascaster, S.M., "Anchor Lines in the North Sea," OTC paper 1535. Presented at the Offshore Technology Conference, Houston, 1972.

12. Childers, M.A., "Mooring Systems for Hostile Waters," *Petroleum Engineer*. 1973.

13. Wilford, F.B., and Anderson, A., "Dynamically Stationed-Drilling SEDCO 445," OTC paper 1882. Presented at the Offshore Technology Conference, Houston, 1973.

14. Duval, B.C. and Corgnet, J-L., "Exploration Drilling on the Canadian Continental Shelf Labrador Sea," OTC paper 2155. Presented at the Offshore Technology Conference, Houston, 1975.

15. Skinner, D., and Hammett, D., "SEDCO 445—Drilling Without Anchors in 2000 Foot Plus Water Depth," OTC paper 2151. Presented at the Offshore Technology Conference, Houston, 1975.

16. "World, Europe Deepwater Drilling Records Set," *Oil & Gas Jour.* (Feb. 7, 1977). p. 34.

17. Bax, J.D., "Drillship Plans use Experience off Labrador," *Oil & Gas Jour.* (Feb. 14, 1977). pp. 119-124.

18. *Deepsea Drilling Project, Technical Report No. 1*. Scripps Institute of Oceanography, San Diego. Oct. 1971. p. 143.

4

Wellheads and Casing

A subsea wellhead, just like a land wellhead, must support the BOPs while drilling, must support the suspended casing while cementing, and must seal off between casing strings during drilling and production operations.

In floating drilling, the casing hangers, casing seals and cementing heads differ from land and platform operations in the following manner:

1. The first and second casing strings are cemented with returns to the seabed.
2. Casing is run with the last joint *made-up* on a casing hanger and permanently suspended prior to cementing. Mud returns flow through fluting in the hanger.
3. Usually, cementing plugs are located at the wellhead and released remotely. The cementing string from the vessel to the wellhead is drillpipe.
4. Casing seals are run and set remotely.
5. Special test tools are required for remotely testing the casing seals.
6. Wear bushings are essential for protecting the wellhead.

Typical casing sizes and depths for subsea casing strings are shown in Table 4-1, with additional data on risers and BOPS. These data are useful when considering BOP and riser sizes and under-reaming requirements. It should be stressed that the nomenclature is not standard for describing casing for subsea completions. The first casing set may be referred to as the *conductor* or the *structural* casing, and the second string may be referred to as either the *surface* casing or the *conductor* pipe. The pipe diameter may also vary, depending on the conditions and the anticipated depth for an exploration well. It will become apparent from the following discussion, however, that the depth to the casing shoe is more important than the size or the name. Each string should be defined by the setting depth and fracture gradient of the formation.

Table 4-1
Sizes of Casing Strings and Holes

Casing String	Casing Size (in.)	Nominal Hole Size (in.)	Coupling O.D. (in.)	Typical Depth Below Seabed (ft)
Structural	30	36	--	100
Conductor	20	26	21.000	1000
Surface	13⅜	17½	14.375	3000
Production	9⅝	12¼	10.625	To prod. zone
Liner or production	7	8¼	7.656	Through prod. zone

OR

Structural	30	36	--	100
Conductor	16	19½	17.000	1000
Surface	10¾	13½	11.750	3000
Production	7⅝	9½	8.500	Through Prod. zone

Wellhead Sizes		Riser Sizes	
Nominal Size (in.)	Drift Dia in.)	Nominal Size (in.)	Typical Drift Dia. (in.)
13 ⅝	12⁵/₁₆	16	14.00
16¾	15⅛	18⅝	16.62
18¾	17½	20	18.00
21¼	19¹¹/₁₆	22	20.00
		24	21.50

The Structural Casing

To get the well started (*spudded*), we need a heavy steel template to guide the bit to the right spot on the ocean floor (see Figure 4-1). The template has four attached guidelines to guide the equipment to the well before the riser has been run. The template is run to the seabed on drillpipe and released mechanically. Then a 36-inch hole opener is run and guided by arms riding on the guidelines so that the bit enters the hole in the template. These are *break away* guide arms that are released by the parting of shear pins or shear bolts after the bit has entered the hole in the template (see Figure 4-2). The

Figure 4-1. Temporary guide base or template.

arms are retrieved to the surface by attached tugger lines. The 36-inch hole is drilled from 80 to 300 feet below the mudline for the 30-inch structural casing.

structured only
no pressure
tolerance

The 30-inch casing is only for structural support, that is, it will not withstand pressure. It is usually drilled and cemented; however, in some areas it may be driven or jetted-in. The casing depth depends on the ability of the soil to support the wellhead and equipment, with the vertical loading and overturning moment used as the criteria for design. If the riser is to be used while drilling the hole for the second casing string, the ability of the formation to withstand the hydrostatic pressure of the mud in the riser must be considered also. Figure 4-3 is a typical plot of allowable mud weight vs water depth for casing set to about 100 feet below the mudline. For the conditions considered, the casing would have to be set deeper than 100 feet in deep water.

The 30-inch casing is run through a permanent guide base with guide posts for the BOPs (see Figure 4-4). These posts are usually hollow, and the guidelines are inserted into the hollow guideposts before running.

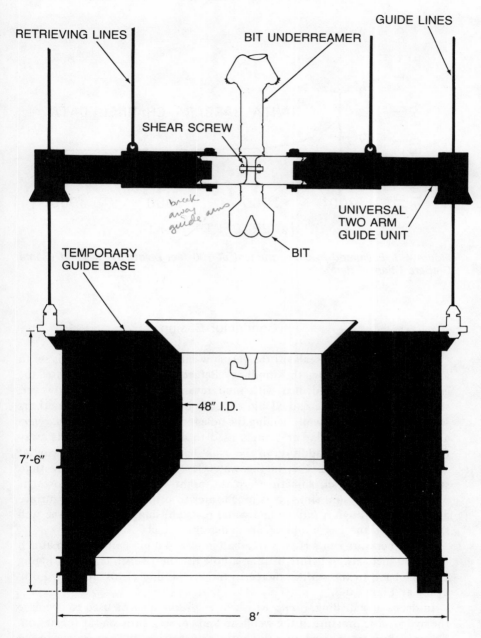

Figure 4-2. Pilot bit being guided into the hole in the template by break away guide arms.

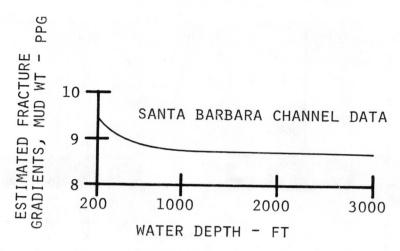

Figure 4-3. *Estimated fracture gradient at 100 feet below the seabed (Santa Barbara Channel data).*

The Conductor Casing

The *conductor pipe* seals off the very low-integrity formations. It is set to about 1,000 feet below the mudline. Before about 1970, holes for the conductor pipe were drilled with mud returns to the seabed. However, shallow gas sands have caused blowouts, and the industry now uses risers with a diverter system while drilling the hole for the 20-inch casing. *Diverters* are low pressure annular preventers used to direct the flow of fluids away from the rig floor. In addition to the usual hazard of fire, a gas blowout at sea can sink a vessel. Gas in the water lowers the density of the fluid supporting the vessel, sometimes to the point where the vessel loses its reserve buoyancy and sinks. It is much safer to bring the gas to the surface and vent it through a *blooey* line while re-establishing control of the well than to have the gas bubbling up under the vessel.

Most operators run a riser with a ball joint and a pin connector to attach to the 30-inch casing before drilling a hole for the 20-inch casing. Bringing cuttings back to the surface increases the probability of lost returns. Look again at Figure 4-3.

In deepwater drilling, *dump valves* or a *lift line* may be used to decrease the hydrostatic pressure at the wellhead and consequently on the formation. Dump valves are located near the bottom of the riser and can be opened to dump the cuttings to the seabed. Dump valves should be two valves in series for redundant closure in the event of a shallow gas kick. A lift line is a line

Figure 4-4. Permanent guide base resting on template.

from the vessel to the bottom of the riser. Water, mud or inert gas can be injected through this line to decrease the hydrostatic pressure.

When a riser is run to drill a hole for the 20-inch casing, a pilot hole is drilled first and then opened to 26 inches. Before the riser is pulled, the mud is conditioned, and a weighted mud is spotted in the well to account for the loss in hydrostatic pressure caused by pulling the riser. The riser has to be pulled because it is too small to accept the 20-inch casing connectors (see Table 4-1). Having pulled the riser, the wellhead and casing are run and cemented with the returns to the sea floor.

While waiting for the cement to set, the riser is run with the BOPs. However, the BOPs should not be closed on the conductor pipe, because at shallow depths there is a strong possibility that the formation fluids and mud may fracture through the sea floor and endanger the vessel.

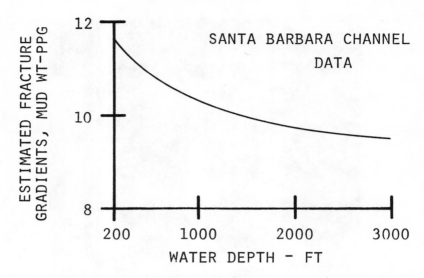

Figure 4-5. Estimated fracture gradient at 1,000 feet below seabed (Santa Barbara Channel data).

The Surface Casing

The *surface casing* seals off the relatively low pressure formations so that the higher pressure formations below 3,000 feet (below the mudline) can be controlled. For the first time since spudding the well, good drilling practices can be used. In this case, *good drilling practices* mean a good mud, good annular mud velocities for hole cleaning, and reasonably good hydraulics. The mud weight and annular mud velocities must be watched because of possible lost returns, especially in deep water (see Figure 4-5).

Drilling the nominal 17½-inch hole for running the 13⅜-inch casing has required appreciable drilling time in the past. One of the reasons for going to wellhead and BOP sizes greater than 16¾-inch has been to avoid the time consumed in under-reaming. Usually over 2,000 feet of hole is drilled for the surface casing. A 17¼-inch bit can be run through an 18¾-inch or larger wellhead, but smaller wellheads and stacks will require that a nominal 17½-inch hole be under-reamed.

By using good floating drilling practices that include proper motion compensation and being aware of the additional pressure caused by the mud in the riser, the hole can be drilled.

The surface casing is run and the casing hanger is hung in the wellhead. After the casing has been cemented, the casing hanger may be jetted with fresh water before the casing seal is run. Two types of seals are commercially available. One type of seal uses drillstring weight to snap locking dogs into prepared groves in the wellhead while energizing the packing. The second type uses torque to set slips into the wall of the wellhead and energize the seal.

Designing the casing program from this point is similar to land wells. The depth of the next casing string will be based on the fracture gradient at the shoe of the surface casing, the estimated formation pressure at the setting depth for the next casing and the estimated pressure gradient of the fluid in the open hole. Pressure integrity and structural integrity are estimated using the same philosophy on land and at sea. The numbers will differ, and they will be discussed below.

Holes for the remaining casing strings are drilled. Each casing is run, cemented and sealed just as the surface casing. Under-reaming is not required for these strings if a $16\frac{3}{4}$-inch or larger stack is used.

Fracture Gradients

Fracture gradients are important in designing a drilling program for offshore operations. The mechanics of hydraulic fracturing of formations has been discussed by various authors, some of which are referenced.[1,2,3,4] Christman[5] applied these techniques to deepwater drilling in the Santa Barbara Channel.

In brief, increasing the water depth reduces the total overburden and consequently the formation fracture gradient. This can be expressed as:

$$g_f = (g_{ob} - g_p) F_\sigma + g_p \qquad (4\text{-}1)$$

where:

g_f = fracture gradient, psi/ft
g_p = formation pressure gradient, psi/ft
g_{ob} = overburden pressure gradient, psi/ft
F_σ = horizontal/vertical stress ratio

For offshore drilling:

$$g_{ob} = \frac{1}{d_{KB}} \left[0.44d + 0.4335\rho_f (d_{KB} - d - d_F) \right] \qquad (4\text{-}2)$$

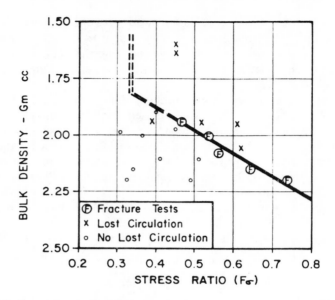

Figure 4-6. Formation bulk density vs. horizontal to vertical stress ratios for the Santa Barbara Channel.

where:

d_{KB} = depth measured from the kelly bushing, ft
d = water depth, ft
d_F = height of flowline above the water, ft
ρ_f = formation bulk density, g/cm^3

Note that $(d_{KB} - d - d_F)$ is merely the penetration into the seabed. Also the effect of pressure caused by the difference in height of the flowline and KB have been neglected.

For the Santa Barbara Channel, the stress ratio (F_o) varied linearly with the bulk formation density (see Figure 4-6). In this case, fracture gradients depend on *soil* characteristics and depths below the seabed but not *water* depth. Christman used formation density logs to determine formation bulk density. Of course, this is a good method for such a determination, and it is recommended by industry. In the event that formation bulk densities are not available, the following application of on-shore fracture data may be used, but with less confidence.

Assume: $g_{ob} = 1.00$ and $g_p = 0.444$

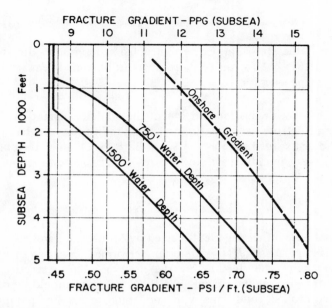

Figure 4-7. An example of onshore and offshore fracture gradients.

by rearranging Equation 4-1 and substituting the assumed values of g_{ob} and g_p, we have:

$$F_\sigma = 1.79 \ g_f - 0.79 \qquad (4\text{-}3)$$

Now using g_f at different depths on shore, we can solve for F_σ at different depths. Often, a stress ratio vs depth correlation will result in a smooth curve that is *nearly* linear (see Figure 4-7).

Now, if we use $0.4335\rho_f = 1.00$, Equation 4-2 becomes:

$$g_{ob} = \frac{1}{d_{KB}} \ [0.44d + (d_{KB} - d - d_F)] \qquad (4\text{-}4)$$

and Equation 4-1 becomes:

$$g_f = (g_{ob} - 0.44) \, F_\sigma + 0.44$$

where:

Fσ is the stress ratio determined from the graph generated from Equation 4-3 for an onshore penetration that is equal to the sea floor penetration offshore.

The problem with this method is that the offshore formation characteristics may differ from the formations onshore. Indications are that in a sandy or consolidated seabed such as the Santa Barbara Channel, the data can be good. In a very muddy sea floor, the results of such estimations may be unreliable.

Leak-off Tests

Details of leak-off testing are discussed by Moore[6] and will be mentioned only briefly here. *Leak-off tests* are often performed on casing that has been set deeper than 1,200 feet below the mudline. This test estimates the formation integrity at the casing shoe. Leak-off tests should be run under carefully controlled conditions with good mud and a low pumping rate. Conventional cementing pumps are marginal in both flow rate and measuring accuracy. Some companies use special low volume pumps for this test.

The test is run after drilling a short distance below the casing shoe (about 10 feet out of the cement). The cement is circulated out, a BOP is closed, and the formation is pressured at a very low rate. The plot of pressure vs. volume in Figure 4-8 is one of the various results that may be obtained from a leak-off test. As soon as the formation begins to take fluid, injection should be stopped: otherwise, the formation will be damaged.

Results are conventionally presented in terms of mud weight required to break down the formation.

$$MW(\text{breakdown}) = \frac{P_s}{Kd} + MW(\text{test})$$

where:

MW (breakdown) = mud weight to breakdown formation, ppg (SG)
MW (test) = mud weight during test, ppg (SG)
P_s = pressure at the surface, psi (Kg/cm²)
d = depth, KB to formation, ft (m)
K = conversion constant, 0.052 psi/ft/ppg (0.10 Kg/cm²/SG)

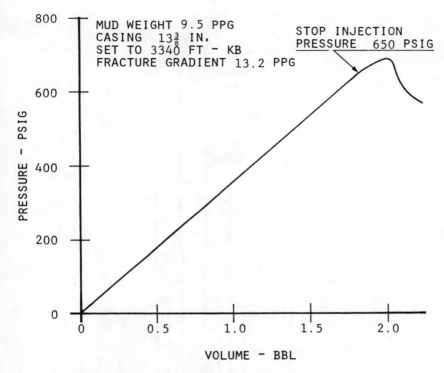

Figure 4-8. Plot of a leak-off test.

Subsea Cementing Systems

In the early wells drilled from a floater, casing extending from the rig to the wellhead was used to land the casing string, and a conventional land-based cementing head was used. Now, unless staged cementing is required, most floaters use subsea cementing systems that hold the plugs in the casing near the hanger (see Figure 4-9). The casing is landed with drillpipe and the plugs are launched by releasing a ball and a dart from the surface. This makes the offshore cementing head much lighter and easier to handle than the conventional cementing heads that are used on land.

The plugs are fitted into the casing, as shown in Figure 4-9, and are retained by shear pins. The *running* tool fits the casing hanger (left-hand thread for *backing-off*) and adapts to the drillpipe. The casing is landed on drillpipe, and connections and other preparations for cementing are carried out just as on land.

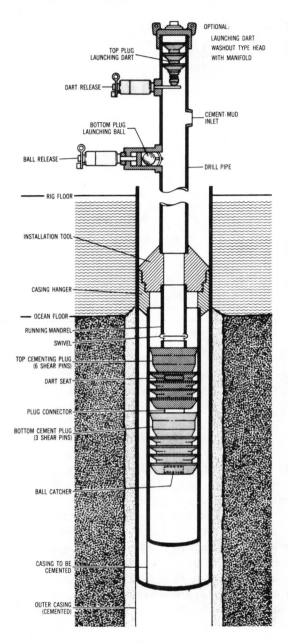

Figure 4-9. Subsea cementing system.

The *lead* (first or bottom) cement plug is launched by releasing a ball from the cementing head. The ball sits in the seating of the ball retainer of the lead plug, and cuts off the flow of circulating fluid until the back pressure creates enough force to shear the pins. The launched plug then acts as if it were on a land well. When the plug bottoms, another pressure buildup will extrude the ball into the ball catcher and allow the cement to flow through the plug.

The *top* (second) plug is launched by releasing a dart. This dart wipes the cement from the drillpipe and launches the plug, just as the ball launched the lead plug. Normal pressure shut-off is achieved when the plug *bumps* (contacts) the lead plug.

Casing Seals

Casing seals are an integral part of the well control system. They must be able to retain well pressure just the same as the BOPs (see Figure 4-10). If a casing seal fails, well pressure will be applied to the weaker, outer casing strings and shallow formations. This will probably rupture the outer casing or cause underground circulation.

The importance of properly setting and testing the casing seals cannot be stressed too much!

Currently, four companies—Cameron Iron Works, National Supply (Armco), Regan International Offshore (Hughes), and Vetco Offshore—manufacture subsea wellhead systems. NL Rig Equipment (Shaffer) is designing a subsea wellhead, but it is not yet on the market.

All subsea wellheads have resilient casing seals to seal off the annulus between casing strings. How the seal is energized and how the seal is held in the energized position is a very important part of the design. All current methods of energizing the seals can be catagorized into three basic types: application of torque to a packing nut (*torque set*), application of weight to the seal assembly (*weight set*) and the application of hydraulic pressure to some form of piston arrangement (*hydraulic set*). The hydraulic set is experimental; to the author's knowledge, it is not in use.

Casing Seal Pressure Tests

Regardless of the wellhead equipment used, verification that an adequate seal has been established is needed prior to drilling out the casing shoe. The tests require careful preparation, and accurate volume and pressure measurements, and even with these precautions the data may be difficult to interpret.

Information needed before the seal is set are the volume of mud in the annulus behind the casing and the volume of fluid (water) from the surface down to the seal. The latter includes the surface manifold, the kill/choke line

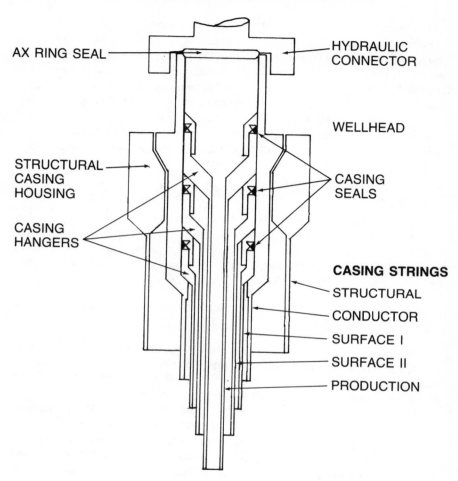

Figure 4-10. Typical sealing arrangement for subsea wells.

to be used, and the volume in the wellhead and stack between the test plug and the closed ram.

The existence of seal leaks can possibly be determined by comparing the volume required to compress the liquid above the seal with the compressibility of the mud behind the seal that would be compressed if the seal leaks.

To adequately discuss casing seal tests, an understanding of liquid compressibility is needed.

Liquid compressibility. We shall discuss theory first. The volume required to compress a liquid is defined by the equation:

$$\Delta V = V_i C_P \Delta P$$

where:

ΔV = increase in volume to increase the pressure by ΔP, vol units
C_P = compressibility of the liquid, vol/vol · pressure
    C_P (water) $\simeq 3 \times 10^{-6}$ psi^{-1} or 2×10^{-7} (kg/cm^2)$^{-1}$
    C_P (mud) $\simeq 6 \times 10^{-6}$ psi^{-1} or 4×10^{-7} (kg/cm^2)$^{-1}$
V_i = system volume, vol units
ΔP = test pressure, psi or kg/cm^2.

Because of the very small values of C_P, ΔV can be expected to be small in a typical field operation.

Water should be used for testing the seal, because it has a lower compressibility, and a less compressible liquid above the seal than below is needed. This accentuates the difference in volumes to be pumped, ΔV, between a good seal and a leak.

Example Situation:

Location of the Cement Top Known
Consider the following conditions:

water depth = 500 ft (all measurements are from the KB)
Casing string = 13 $^3/_8$ in. set at 4,000 ft
volume of system above seal = 11 bbl
test pressure = 3,000 psi
test fluid = water
previous casing string = 20-in., J-55, 94.0 lb/ft, set to 1,500 ft
cement top = 996 ft

With no leak, the system will require

$$\Delta V = 3 \times 10^{-6} \times 11 \times 3,000 = 0.1 \text{ bbl } water$$

to reach test pressure.

If the seal leaks, the volume will be more, but how much more? Obviously, 0.1 bbl would be difficult to measure. The annular volume between the seal and the cement is

$$(996 - 500) \text{ ft} \times 0.1815 \text{ bbl/ft} = 90 \text{ bbl of } mud$$

now,

$$\Delta V = 6 \times 10^{-6} \times 90 \, \Delta P + 0.1 = 5.4 \times 10^{-4} \Delta P + 0.1 \text{ bbl}$$

Pressure in the annulus must always be less than the collapse pressure of the inner casing, and less than the internal yield of the outer casing. This will depend on both volume and pressure. Table 4-2 shows the relationship for four grades of casing. Also, the internal yield of the 20-inch casing is reached at 2110 psi when $V = 1.22$ bbl.

Table 4-2
Relationship of Casing Grades

13³⁄₈-in. Grade Weight per Ft (lb/ft)	Collapse Pressure (psi)	ΔV to Reach Collapse Pressure (bbl)
J–55/54.5	1130	0.7
J–55/61.0	1540	0.9
J–55/68.0	1950	1.2
N–80/72.0	2670	1.5

Table 4-2 indicates that small volumes may exist between the volume required for a seal test and the volume that would collapse the casing, especially where light casing is used. Seal tests should be taken into account when casing is being designed.

So much for the theory: Now back to the real world.

Even with water as a test fluid, more than 0.1 bbl will probably be pumped to pressure the 11 barrels above the seal from 0 to 3,000 psi. The errors result from inaccurate measurements, variations in C_p, air in the lines, and the expansion of the system—especially the flexible lines.

It is evident that there is no room for sloppiness when running a casing seal test. Some judgment may be required to determine if a seal is leaking, even when the utmost care has been used to run the test.

General procedure. After setting the seal, a test plug is landed in the wellhead, sealing off the casing below the seal (see Figure 4-11). The plug may be designed to seal a short distance into casing. In this case, the test pressure must be below the internal yield pressure of the casing. Another type of test plug seals right at the casing hanger and does not subject the casing to test pressure. For the latter case, the seal can be tested to the full working pressure of the wellhead.

After the test plug has been landed, a ram is closed and the mud is displaced with water. Then the test pressure is applied above the test plug. The test pressure and volume of fluid pumped is monitored and recorded. Applying the pressure in stages will reduce the probability of collapsing the casing if the seal does leak.

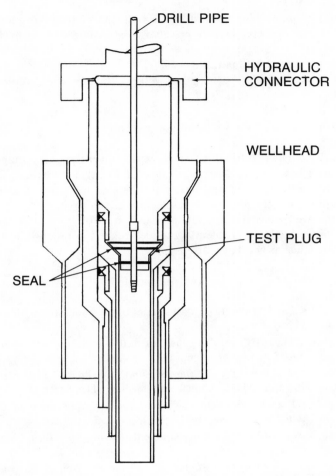

Figure 4-11. Plug for testing casing seal to full working pressure.

Test pressure. There are two schools of thought on what level of test pressure should be used for testing casing seals. One school holds that rigorous tests should be made: The seal should be tested to the full working pressure of the wellhead. The second school maintains that testing beyond the burst pressure of the casing is unnecessary, because the casing will burst before the wellhead pressure can be applied to the seal. Proponents of this school also point out that pressures higher than necessary will increase the probability of collapsing the casing if the seal does leak.

Test Evaluation. During the test, if the wellhead system being tested will not sustain test pressure, several possible causes should be considered:

1. Leak in the surface manifold
2. Leak in the test plug (detected by returns through the drillpipe)
3. Leak in the casing seal
4. Leak in the BOPs
5. Leak in the hydraulic wellhead connector

The last three items are difficult to isolate with conventional test plugs. Some manufacturers have special test plugs that seal directly above the casing seal. This type of tool is isolated and tests only the seal assembly. These special test tools may or may not be available in a given area.

When the well does not sustain pressure, it is obvious that there is a problem. There is also a problem if the well takes too much fluid to reach test pressure, just as we have discussed.

References

1. Hubbert, M.K., and Willis, D.G., "Mechanics of Hydraulic Fracturing," *Trans. SIME.* p. 153.
2. Dunlap, I.R., "Factors Controlling the Orientation and Direction of Hydraulic Fractures," *Jour. Inst. Patrol.* (Sept. 1963). p. 282.
3. Deily, F.H., and Owens, T.C., "Stress Around a Wellbore," SPE paper 2557. Presented at the SPE-AIME Annual Fall Meeting, Denver, 1969.
4. Eaton, B.A., "Fracture Gradient Prediction and Its Application in Oil Field Operations," *Jour. Petro. Tech.* (Oct. 1969). p. 1353.
5. Christman, S.A., "Offshore Fracture Gradients," *Jour. Petr. Tech.* (Aug. 1973). p. 910.
6. Moore, P.L., *Drilling Practices Manual.* Petroleum Publishing Co., Tulsa, 1974.

Blowout Preventers
and Their Control

The blowout preventers (BOPs) are designed to shut in a well under pressure so that formation fluids that have moved into the well bore can be circulated out while continuous control of the well is maintained. More than one type of BOP is used and, for redundancy, two or more preventers of the same type are used. These will be stacked together and, with the appendages to be explained later, the assembly is referred to as a BOP stack, or simply the *stack*.

The Stack

Various laws pertaining to BOP requirements were revised during 1976 and 1977 by different countries. Laws of the countries where rigs will be drilling should be studied carefully regarding BOP requirements.

Technology and state-of-the-art design and use of BOPs on land were well established at the advent of floating drilling; however, it became apparent that when BOPs are located on the ocean floor, considerable changes were required. These changes were not only on the stack itself; the control system, the drilling procedure and the procedure for using the equipment also had to be changed. Major changes in the equipment for subsea use are:

1. The size of the BOPs is increased.
2. External hydrostatic pressure at the ocean floor must be considered.
3. Hydraulics have become more important in reaction times, because the longer flow lines increase the pressure drop, while the larger BOPs require more fluid to operate than their land counterparts.
4. To avoid the pressure drop in return lines, the hydraulic fluid is vented to the sea. This requires a fluid that is non-polluting as well as non-corrosive, yet has low viscosity, is a good lubricant and can mix with water of high mineral content.
5. The philosophy of stack arrangement, especially the location of the kill and choke lines has been changed.

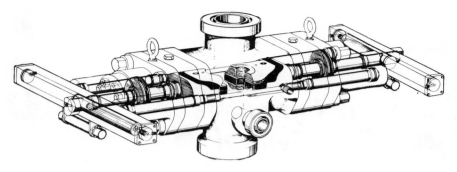

Figure 5-1. Cameron U-type ram blowout preventer.

    6. Pressure drop in the long choke line(s) influences well control pro-
cedures.

Present BOP design and arrangement came about through a pragmatic
evolution, not through a sweeping scientific reform. By late 1973, major
shortcomings of the equipment had been corrected, so some chronology will
be injected into the logical order of presentation here. Here, we will discuss
the BOP equipment, then the arrangement, and finally the control system.

Rams

The massive steel rams have rubber seals, and are hydraulically actuated
to seal off the well bore (see Figures 5-1 and 5-2). Pipe *rams* seal the annulus
around the drillpipe and are designed so that an entire string of drillpipe and
collars may be *hung-off* (suspended) from a pipe joint landed on a ram. The
ram seals must be the correct size in order to seal. In other words, 3-inch
seals cannot be used for 5-inch drillpipe. Conventionally, three pipe rams are
used. A fourth ram, a shear-blind ram, is used to seal over the open hole and
to shear the drillpipe when necessary. Shearing the pipe is of course one of
the last resorts in an emergency situation.

There are two major manufacturers of BOP rams: Cameron Iron Works
and NL Rig Equipment (Shaffer). At the 1976 Offshore Technology
Conference, the Hydril rams were presented. At the writing of this book,
these rams had not been used subsea. Consequently, this discussion will be
limited to the Cameron and Shaffer rams.

Five specifications are considered important for subsea ram operations.
Current models of both major manufacturers meet these criteria. Earlier
(pre-1973) models will meet some but not necessarily all of these specifications:

    1. The BOPs must be able to sustain the internal pressure stamped on the
BOP case.

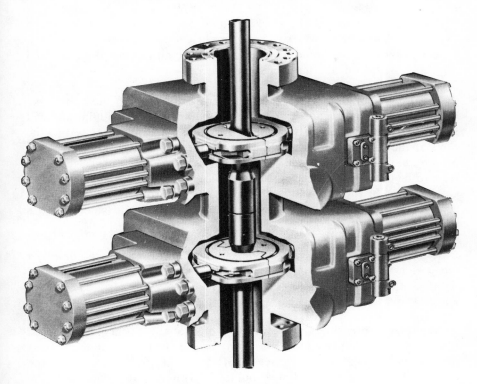

Figure 5-2. Shaffer LWS-type blowout preventers.

2. The pipe rams should be able to suspend a drill string weighing up to 600,000 lbs on a tool joint and still maintain the specified working pressure. Disconnecting on the tool joint may be advisable when circulating out kick, when pipe has to be sheared, and for drill-stem testing. One important requirement is that the ram blocks be harder than the pipe tool joint. This allows the rams to cut into the joint instead of allowing the tapered joint to act as a wedge to pry open the rams.

3. The pipe rams should be able to withstand pressure imposed above the ram. This is needed when returning to the well after hanging-off, disconnecting the drill string and closing the shear/blind rams. When returning to operations, the well is circulated by pumping in through a kill/choke line above the ram, down the pipe, up the annulus and out a kill/choke line below the ram. Current equipment will withstand some pressure, and full working pressure is not required above a ram.

4. The ram packing should be able to withstand seawater pressure when

internal pressure of the stack is about one atmosphere. This can be a problem in deep water when severe lost returns are encountered.
5. The shear rams should be able to shear the pipe and seal off the well even if stubs of the pipe remain in place.

Pipe rams should be closed only around pipe, never on the open hole, because the sealing surface is designed to seal around the drillpipe and will be extruded and worn excessively if the drillpipe is absent. When testing the rams at the surface, special precautions should be taken to assure that the *primary* seals are being tested. Both manufacturers have plastic secondary seals to be used in emergencies. Secondary seals are actuated by divers and are energized by rotating a screw that injects plastic around a ram shaft. This is a *static* seal: it wears rapidly as the ram is operated and will accelerate shaft wear.

Secondary seals should be used only in emergency and should be taken out of service as soon as practicable aften the emergency is over.

Annular Preventers

The annular preventers are comprised of specially designed, reinforced rubber elements that can seal around any tubular or nearly tubular objects that will go through the BOPs (see Figure 5-3). They will also seal over the open hole, and can pass drillpipe tool joints without severe damage to the sealing element (see Figure 5-4). Because of their flexibility, the annular preventers are often referred to as *universal preventers*. Annular preventers are actuated by an annular piston that squeezes the seal into the bore. The piston area is large relative to the other functions on the stack and, except for initial closure, should be operated at pressures lower than the other functions on the stack. This decreases the possibility of extruding the rubber seal out of the preventer.

Annular preventers are used for stripping drillpipe into the well under pressure and for shutting in the well around tubing when the pipe rams cannot be used. In particular, annular preventers are used for initial well shut-in during a kick and to hold the well pressure until a drillpipe tool joint is located and hung-off on a pipe ram. When rams are closed, the position of the tool joint near the ram must be known, because a heave of the drilling vessel at the wrong time could cause extensive and costly damage to the equipment.

Frequently two annular preventers will be used. One of these preventers will normally be located above the upper hydraulic connector so that it can be retrieved with the riser.

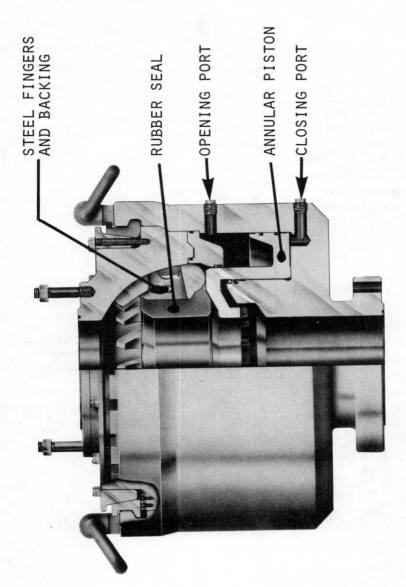

STEEL FINGERS AND BACKING

RUBBER SEAL

OPENING PORT

ANNULAR PISTON

CLOSING PORT

Figure 5-3. Annular blowout preventer.

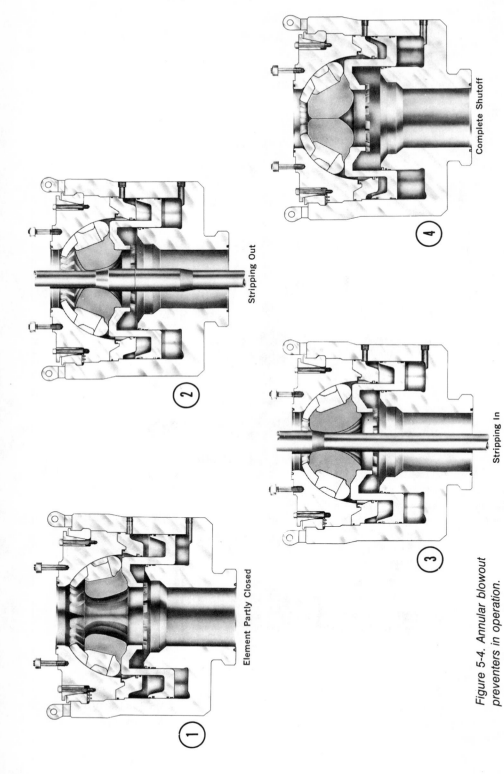

Element Partly Closed

Stripping Out

Stripping In

Complete Shutoff

Figure 5-4. Annular blowout preventers in operation.

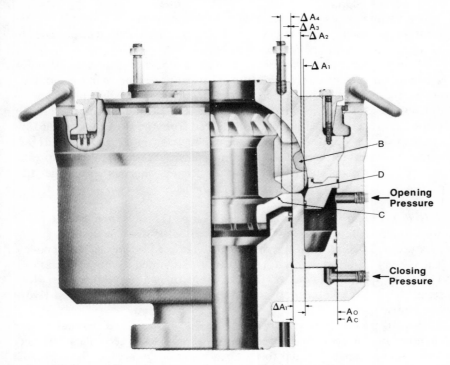

Figure 5-5. Shaffer spherical blowout preventer—cutaway.

Shaffer annular preventer. The *Shaffer Spherical* can be closed at 1,500 psi on any diameter pipe without damaging the rubber element. However, lower pressures are recommended for stripping. When the element is fully opened, seals B and C are inactive, allowing pressure to equalize around the element (see Figure 5-5). The pressure in the mud column acts on Area A_1 and is opposed by the subsea pressure on the close side. The net result created by the difference in pressure of the mud column and the water is usually a downward force tending to hold the preventer open.

When closing pressure is applied, the piston moves upward and seals B and C are activated. A_1 then shifts to A_2. The right side of A_2 is the top edge of seal B. As the piston moves upward, area A_2 becomes smaller until it reaches zero at A_3.

Less closing pressure is required to seal under water than on land because the element seals on the pipe after seal B is in area A_4. Area D extends upward to the top edge of seal B, and a very low pressure exists between the element and the upper housing, because area D becomes larger as the piston strokes upward. When the seal is closed, the well pressure acts on area A_4 to increase the closing force on the piston.

When stripping, the pressures need to be decreased below 1,500 psi to avoid damaging the sealing element. Typical closing pressure curves recommended by Shaffer are shown in Figure 5-6.

Shaffer recommends the use of a surge chamber (5-gallon or 10-gallon accumulator) on the closing side of a spherical preventer. When a tool joint enters the element, the rubber is forced outward and the piston moves downward. Both the rubber element and the hydraulic fluid are nearly incompressible, and hydraulic fluid is forced from the closing side. The flow rate and volume of this fluid cannot be handled adequately by a regulator, so surge chambers are needed. In an accumulator, the gas merely compresses and expands to accept and redeliver fluid. Shaffer recommends a precharge of 50 psi per 100 feet of water depth for the accumulator.

Hydril annular preventer. The Hydril Company makes three basic types of annular preventers. The low pressure rated MSP has a maximum service pressure of 2,000 psi. This version is not often used on subsea stacks today. The GK Hydril was designed for land, but was adopted to subsea operations. The large subsea versions of the GK went through modifications which resulted in the development of the GL preventer that is the preferred Hydril for subsea use.

The GL preventer is shown in Figure 5-7. It has equal open and close chamber piston areas and a *secondary chamber*. The secondary chamber area is approximately equal to the area acted on by mud in the riser. The hydrostatic pressure in the riser tends to hold the preventer open, and the secondary chamber can be used to overcome this force.

There are three possible combinations for hooking up the secondary chamber (see Figure 5-8). One method is to connect the secondary chamber to the open chamber. When closing pressure is applied, the difference in pressure required to close the Hydril is the difference in pressure between the mud in the riser and the ambient sea pressure.

The second method is to connect the secondary chamber with the closing chamber. This configuration increases the closing force, above the closing force of the well (and above normal close pressure), and may be undesirable when stripping.

The third method is to connect the secondary chamber to the marine riser through an accumulator. Now the secondary chamber becomes a *balance chamber* to compensate for the mud column pressure that tends to hold the element open. With this configuration, the closing pressure will be independent of the water depth. However:

1. Any hole put in the riser should be accompanied by a detailed study of its effect on riser integrity.
2. Mud solids have a tendency to plug such an installation.

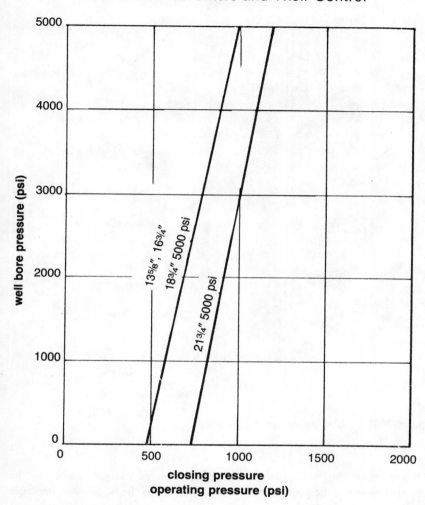

Figure 5-6. Typical wellbore pressure vs. closing pressure curves, Shaffer spherical BOP with 5″ D.P. (Actual values may vary. See operating manual for specific preventer and pipe size.)

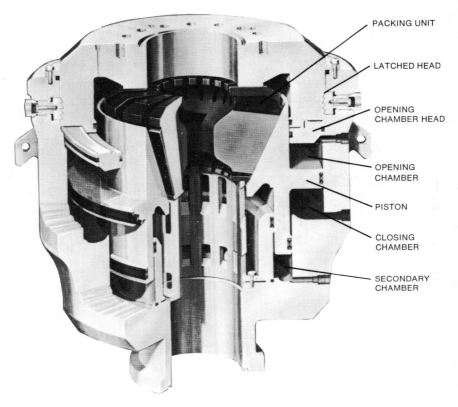

PACKING UNIT

LATCHED HEAD

OPENING
CHAMBER HEAD

OPENING
CHAMBER

PISTON

CLOSING
CHAMBER

SECONDARY
CHAMBER

Figure 5-7. The GL Hydril blowout preventer.

Of the three hookups shown, connecting the secondary chamber to the open port seems to be preferred by industry because the additional pressure required to close the Hydril is only the difference in the hydrostatic pressure of the mud column and sea water.

Surge chambers are recommended by the manufacturer for both the open and close ports. The recommended precharge pressures are:

Closing chamber
$$\text{precharge} = 0.8 \ [600 + (0.41) \ (\text{water depth, ft})]$$

Opening chamber
$$\text{precharge} = 0.8 \ (0.45) \ (\text{water depth, ft})$$

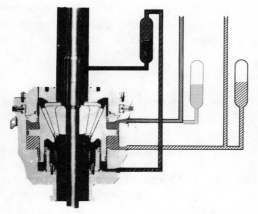

SECONDARY CHAMBER
TO RISER

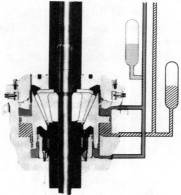

SECONDARY CHAMBER
TO OPEN CHAMBER

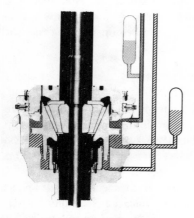

SECONDARY CHAMBER
TO CLOSE CHAMBER

Figure 5-8. Secondary chamber arrangements for the GL Hydril.

The operator's manual for the GL Hydril give a series of tables and nomographs for determining the proper closing pressure under essentially all anticipated conditions with the exception of stripping.

Hydraulic Connectors

Hydraulic connectors are used between the wellhead and the BOPs, and between the BOPs and the riser. These connectors are controlled from the surface and there are two basic types.

The *mandrel-type connectors* secure locking dogs around flanges that are machined into the wellhead. Manufacturers of mandrel connectors are Vetco Offshore, Regan International (Hughes) and National Supply. The dogs are driven by cams that are actuated by hydraulic pistons (see Figure 5-9).

The *collet connector* is used on wellheads designed with a Cameron wellhead hub. This connector uses a series of collet fingers that form a funnel configuration to guide the connector over the well hub (see Figure 5-10). The collet fingers are driven by a ring of cams that are actuated by hydraulic pistons. The collet connector is manufactured by Cameron Iron Works.

Both types of connectors use AX-ring seals. Manufacturers make available AX-rings with resilient coating. This ring can seal at times when the all-metal ring will not seal. There are various reasons for a resilient seal's working when an all-metal seal will not. For example, a steel wire that has cut into the wellhead is a situation often impossible to diagnose from a floater. Under such circumstances, a sense of false security caused by a successful pressure test can lead to a hazardous situation. Therefore, the author recommends that a resilient seal *not* be used in day-to-day operations. In fact, resilient seals in the connector should be considered a risk for any operation.

Kill and Choke (K&C) Valves

Kill and choke valves are the subsea shut off of the high pressure (kill/choke) lines that run from the BOPs to the choke manifold on the rig. The kill/choke lines on the riser are an integral part of the riser and will be discussed in Chapter 6; however, the K&C valves are a part of the stack, and are controlled by the BOP control system.

K&C valves are hydraulically controlled from the surface and are designed to close by spring action when opening pressure is released (see Figure 5-11). A valve may be closed hydraulically or it may use the *fail-close* feature for closing the valve in normal operations.

Two valves in each line should always be used for redundancy. The valves should be located as close as possible to the ram body for mechanical protection. Some of the older valves may be directional and seal from only

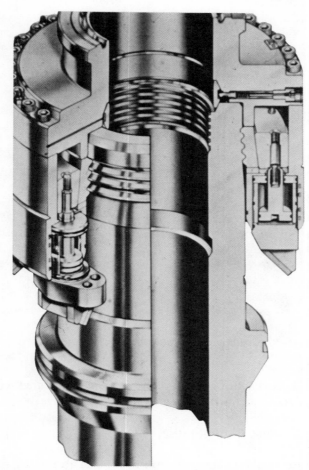

Figure 5-9. Mandril type hydraulic connector.

one direction, depending on the design. Only one unidirectional valve should be used per line. The unidirectional valve should be the valve closer to the BOP, and must be installed to seal against internal BOP pressure. This valve is used to avoid hydraulic blocks that may occur in some valve systems when both valves are closed with liquid under pressure in the line. Some valves do not have a problem with reopening, regardless of the pressure in the line when both valves are closed; therefore, care should be taken to select the proper K&C valves to avoid these problems.

K&C valves must be rated for the same working pressure as the BOP stack. Kill/choke lines located on the stack should have targets in all 90°

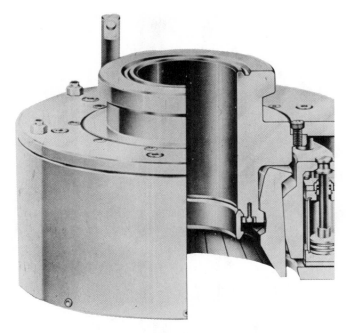

Figure 5-10. Collet type hydraulic connector.

bends to avoid sand cutting. K&C valves are manufactured by WKM, McEvoy, Shaffer, Cameron Iron Works and Vetco Valves.

Philosophy of Stack Configuration

To paraphrase an old saying, you never miss being able to see, repair and adjust the BOPs until they become inaccessible. Subsea BOPs require additional redundancy that is unnecessary for surface stacks, and various combinations and configurations have been tried. Some have proved to be more suitable than others. Industry trends appear to be toward a single stack system (as opposed to a two stack system) where the shallow hole is drilled with minimal low pressure equipment and the lower formations are drilled with a smaller, higher pressure stack. The trend may be reversed. Industry may be forced into the two stack system if very high (greater than 10,000 psi) pressures are to be encountered in floating drilling. Popular sizes for single stacks are 16¾, 18¾ and 21¼.

The stack configuration presently accepted by industry for the major components is shown in Figure 5-12. From the connector up, it consists of a wellhead connector, three pipe rams, a set of shear/blind rams, an annular

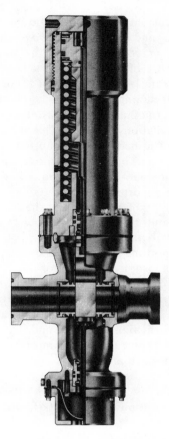

Figure 5-11. Fail-safe kill and choke valve.

preventer, the riser connector and a second annular preventer. However, the location of kill/choke line outlets on the stack vary considerably between rigs and are the major configuration differences in single stack systems used currently.

Designating one line as the *kill line*, and one as the *choke line* is a carryover from land operations. It is inappropriate on a floater. On surface BOPs, two lines branch from a cross just above the bottom pipe rams. One line goes to the choke manifold, and the other line goes to a kill manifold for ganging pumps to inject into the annulus to kill the well. On a floater, both primary lines from the surface can be manifolded to pumps or to the chokes. In floating drilling, there is the additional need to be able to hang-off and disconnect, as previously discussed. Before the shear/blind rams are reopened, the well is circulated by injecting mud down one line, through the drillpipe, up the annulus and out the other *choke* line. That same *choke* line may be used on another occasion to *kill* the well by injecting into the annulus, the sole purpose of a *kill* line on a land rig.

Another carryover from land drilling, still followed by some but not all operators, is to avoid installing a line below the bottom rams. On land rigs, there is a problem with roughnecks and roustabouts opening valves or even removing them for installation elsewhere. This is not a problem with a stack on a subsea wellhead. An incident was reported in the North Sea area in which an anchor caught and broke a kill/choke line below the bottom pipe rams, resulting in a blowout. To the author's knowledge, this is the only incidence of this type reported. Evidently there is a *risk*, but with the apparently low probability of recurrence of the North Sea incident and the additional flexibility and redundancy offered, various operators prefer having a line below the bottom rams on subsea stacks. On a subsea stack, only rams above the bottom kill/choke line can be tested.

To appreciate the philosophy of choice of outlets for the stack, example configurations of kill/choke line outlets are shown in Figure 5-12, and a comparison of the flexibility afforded by the various configurations are shown in Table 5-1. The table is based on assumptions that the ram configuration shown in the figure is used, and that while up to four outlets will be allowed between the rams, some must be interconnected, because only two high pressure lines are available on the riser.

This is an example of the philosophy used in selecting or modifying existing equipment and must be combined with the operating procedures of the operator. All configurations shown in the figure were in field use by mid-1977.

BOP Controls

Control systems for the BOPs are of necessity highly efficient hydraulic systems. The objective is to operate any of the functions on the stack in as short a time as possible. This requires high flow rates for the large volumes needed to operate the major functions such as the rams and annular preventers. For example, eight seconds is an acceptable closing time for a ram. Volumes required for closing preventers varies from about five gallons for small rams to over fifty gallons for a large annular preventer.

Two basic types of control systems are presently used: the hydraulic and the electrohydraulic systems. There are three major manufacturers: NL Rig Equipment (Stewart & Stevenson Oil Tools), Payne Products Division of Cameron Iron Works, and Valve Control Company (Hydril).

During this discussion, the NL (Koomey) system will primarily be used to illustrate the concept of redundant hydraulic operation. The equipment

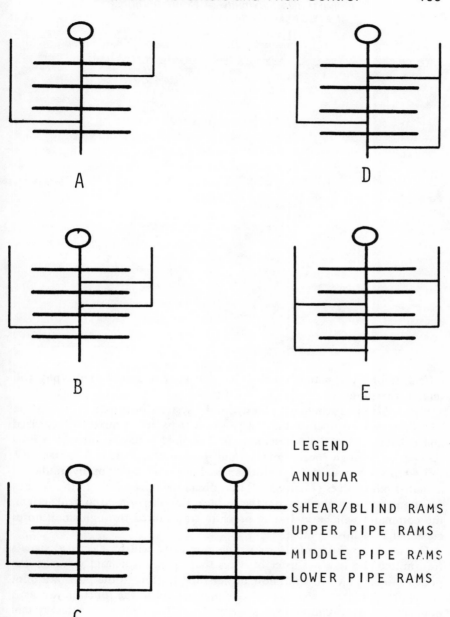

Figure 5-12. Example configurations of kill/choke line outlets for conventional four-ram stack.

Table 5-1
Comparative Flexibility of Configurations of Kill/Choke Outlest (see Figure 5-12)

CONFIGURATION	NORMAL CIRCULATION			DRILLSTRING DISCONNECTED		
	LINE	NUMBER OF HANG-OFF POINTS	RAMS	LINES	NUMBER OF HANG-OFF POINTS	RAMS
A	T	0	ANNULAR ONLY	T/B	2	UPR/MPR
	B	2	UPR/MPR			
B	T	0	ANNULAR ONLY	T/B	2	UPR/MPR
	M	1	UPR	M/B	1	MPR
	B	2	UPR/MPR			
C	T	1	UPR	T/M	1	MPR
	M	2	UPR/MPR	M/B	1	LPR
	B	3	ALL			
D	T	0	ANNULAR ONLY	T/M	2	UPR/MPR
	M	2	UPR/MPR	M/B	1	LPR
	B	3	ALL			
E	T	0	ANNULAR ONLY	T/UM	1	UPR
	UM	1	UPR	UM/LM	1	MPR
	LM	2	UPR/MPR	LM/B	1	LPR
	B	3	ALL	T/B	3	ALL

ABBREVIATIONS:

LINES
T - TOP LINE
M - MIDDLE
UM - UPPER MIDDLE
LM - LOWER MIDDLE
B - BOTTOM

RAMS
UPR - UPPER (MOST) PIPE RAM
MPR - MIDDLE PIPE RAM
LPR - LOWER (MOST) PIPE RAM

differs among the manufacturers, but the system design philosophy will remain the same.

Figure 5-13 is a schematic of a hydraulic system for subsea BOP control. The fluid used to operate the functions on the stack is mixed, pressurized and delivered from the *hydraulic unit*. The fluid is passed through a *hose bundle*, or through special power fluid lines on the riser, to a subsea *pod*.

Two pods are used for redundancy. Each pod contains hydraulically actuated pilot valves (Figure 5-14) that direct the power fluid to the various functions upon command from the surface. For a *hydraulic* control system, the command will be hydraulic pressure transmitted through small hoses clustered in a hose bundle.

For an *electrohydraulic* system, the command will be electrical signals transmitted through a cluster of wires that operate solenoid valves in the pod. A solenoid valve will allow power fluid to flow to the port of a pilot valve, actuating the valve that directs power fluid to the function just as is done in the hydraulic control system. Thus, the electrohydraulic system operates similarly to the hydraulic system, with the exception of the nature of the commands from the surface and of the solenoid valves in the pods.

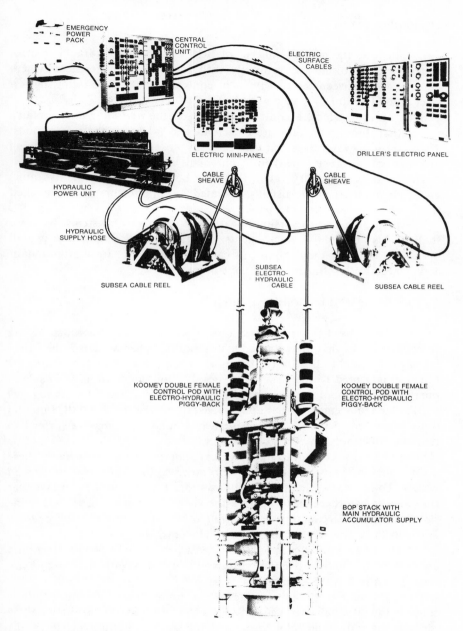

Figure 5-13. Schematic of a hydraulic control system.

Hydraulic Control

To illustrate the general concept of BOP control, the hydraulic control system will be discussed, primarily with reference to differences (where appropriate) between the two types of control systems. The following is a brief simplification:

Back on the surface, a control manifold on the *hydraulic unit* contains valves that direct pilot pressure to actuate the valves on the subsea pods. These are four-way valves that vent one (say close) pilot line while applying pressure to the other (open) pilot line. This avoids a hydraulic block.

These valves are rotary valves that may be operated from the hydraulic units. However, they are usually operated by air pressure and controlled from one of several remote panels.

The preceding is an oversimplification of the control systems. Now we shall discuss the major components. Next, we shall discuss the overall operation and how the components can be combined to form a redundant and hydraulically efficient BOP control system.

Components of a Hydraulic System

Pilot valves used to control the flow of power fluid to the functions are three-way valves that are actuated hydraulically. When actuated, the valves allow power fluid to flow to the function. When not actuated, the line to the function is vented to the sea. Two types of pilot valves are used in subsea operations. They are the *poppit valve* (see Figure 5-14) and the *sliding seal valve*. One requirement is that upon valve failure, the line from the function must be vented to the sea; otherwise, a hydraulic block may be formed so that the function cannot be carried out.

Pilot valves are controlled by individual *pilot lines* from the surface. These are 1/4-inch or 5/16-inch ID lines, reinforced so that very small expansion occurs. To actuate a function, fluid flows into the line until the pressure at the valve is high enough to operate the valve. Expansion of the pilot line slows down the reaction time. Power *fluid* is furnished through a large *power line* usually located at the center of a bundle of smaller pilot lines (see Figure 5-15). The power fluid line should be at least 1-inch ID: larger diameter lines are preferred. The objective is to move as *much fluid* as possible in *minimum time* with the available *operating pressure*.

Thus, the control system operates in an indirect manner. Power fluid is ready at the pilot valves on the subsea stack. The pilot valves are actuated by pressurizing individual pilot lines from the surface. Fluid expelled by the function is vented.

Increasing the water depth requires increasing both power and pilot line lengths, resulting in increased reaction times of the functions (see "Factors

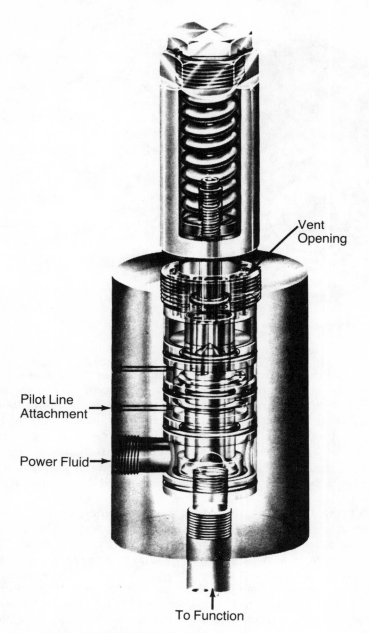

Figure 5-14. Pilot valve—Poppet type.

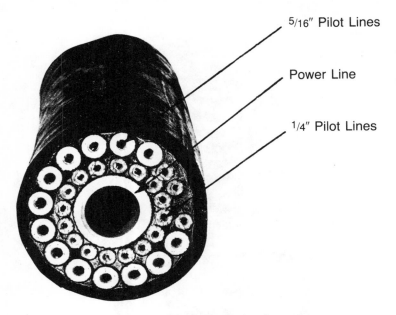

5/16" Pilot Lines

Power Line

1/4" Pilot Lines

Figure 5-15. Typical hose bundle for hydraulic system.

Influencing Reaction Times" later in this chapter). Electrohydraulic systems can decrease reaction times in deepwater operations. Instead of running pilot lines from the surface, pilot lines are run from the power fluid line in the pod on the stack through an electronic solenoid valve and then to the pilot valve. The signal from the surface travels at approximately the speed of light through the connecting wires. Only a very short steel pilot line is used to operate the pilot valve. This decreases the reaction times appreciably in deep water. Figure 5-16 illustrates a typical electrohydraulic hose bundle.

With hydraulic systems, a wire rope is used to retrieve the pod separately from the risers and BOPs. This wire rope also supports the weight of the hose bundle that is attached to the wire rope by special clamps. At the surface, tension is held on the wire rope, and the hose bundle requires support only from the top clamp. In the electrohydraulic systems, steel strength members (No. 4 in Figure 5-16) are included internally, allowing the entire bundle to be tensioned. This eliminates the requirement for the wire rope and the worries of connecting it to the bundle. It should be mentioned, however, that tensioned lines deteriorate at the point where they are reeved for tensioning. This is a point to consider before deciding whether it is economically expedient to eliminate the wire rope and use the internal strength member to support the bundle.

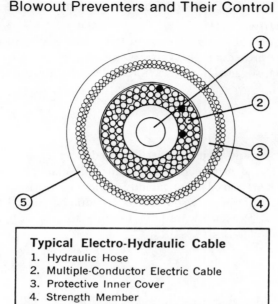

Typical Electro-Hydraulic Cable
1. Hydraulic Hose
2. Multiple-Conductor Electric Cable
3. Protective Inner Cover
4. Strength Member
5. Outer Protective Jacket (Optional)

Figure 5-16. Typical hose bundle for electrohydraulic system.

A system of the near future is the multiplexed electrohydraulic system. This system requires only two wires to carry all signals to and from each pod. More than two are used for redundancy. The signals are coded with a name tag of the function to be operated. On the pod, a logic circuit reads the signal and operates the proper function via the solenoid and pilot valves.

With the multiplex system, the cost and problems of large lines to the pods are decreased by a sophistication of the subsea electronics. Multiplex systems hold the promise of decreasing the operational problems of running and retrieving the lines to the pods, but reliability of the system must be demonstrated before it will be accepted for a primary control system.

Regulators

Power fluid is furnished to the control pod at the working pressure of the system, but the stack functions are designed to work at lower pressures, requiring pressure regulation on the pod. These regulators are controlled from the surface. The *regulator* is essentially a sliding seal valve that balances a *pilot* (control) pressure against the pressure on the output side of the regulator. If the output pressure drops below the pilot pressure, the valve moves to allow power fluid to increase the pressure. If the output pressure is higher than the pilot pressure, the valve slides in the other direction, equalizing the output and pilot pressures by venting. When the output and pilot pressures are equal, the valve is centralized to shut off flow. For regulator sizing, see section on "Factors Influencing Reaction Times," later in this chapter.

Control Pods

The pilot valves and regulators are clustered together in *control pods* (see Figure 5-17). Two control pods are used for redundancy and, on most rigs, either pod can be retrieved independently of the stack or riser. The pods have a guiding and orienting mechanism and a hydraulic latch that is operated from the surface. Proper orientation is essential for proper operation of the stack.

Accumulators

To have the necessary quantity of control fluid available under pressure requires storing this fluid in *accumulators* (see Figure 5-18). These accumulators operate by the expansion and compression of nitrogen gas that is separated from the hydraulic fluid by either rubber bladders or pistons. The hydraulic fluid is pumped into banks of these accumulators that are manifolded together.

Accumulators are used both at the surface and on the stack. The nitrogen charge, fluid capacity and total accumulator capacity for a rig are discussed under "Accumulator Capacity" later in this chapter.

Shuttle Valves

One of the simplest yet very important pieces of equipment on the subsea control system is the shuttle valve (see Figure 5-19). For redundancy, two control pods are used. The *shuttle valve* diverts the flow from the active pod to the function and isolates the inactive pod. Thus, the lines from the two separate pods are redundant up to the shuttle valve, but between the shuttle valve and the function there is no redundancy. Consequently, the shuttle valve should be located as closely as possible to the function.

Lines on the Stack

Lines to the major functions on the stack should be as short as possible and have as big an inside diameter as practicable. A minimum of ells and tees should be used in the plumbing. Good hydraulics are required for fast reaction times and will be discussed (see "Factors Influencing Reaction Times," later in this chapter).

Figure 5-17. BOP control pod.

Hydraulic Unit

The *hydraulic unit* is the power source for the control system (see Figure 5-20). It consists of:

1. An atmospheric storage tank for control fluid.
2. Pressure pumps to bring the control fluid up to accumulator pressure.
3. Accumulator banks for storage of the pressurized fluid.

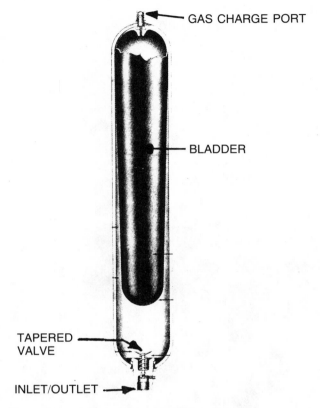

Figure 5-18. Accumulator, bladder type.

4. Rotary valves that pressure and vent pilot lines to open or close each function.
5. Air pistons to operate the rotary valves from a remote location.
6. A selector valve to direct the flow of either power or pilot fluid to the active pod and isolate the inactive pod. This valve may also be operated remotely.

Air must be supplied to control rotary valve operation. The air supply, however, is not usually included with the power unit.

Redundant Operation

Now that we have discussed the components of a subsea BOP control system, it is time to discuss how these components are combined into an operating system.

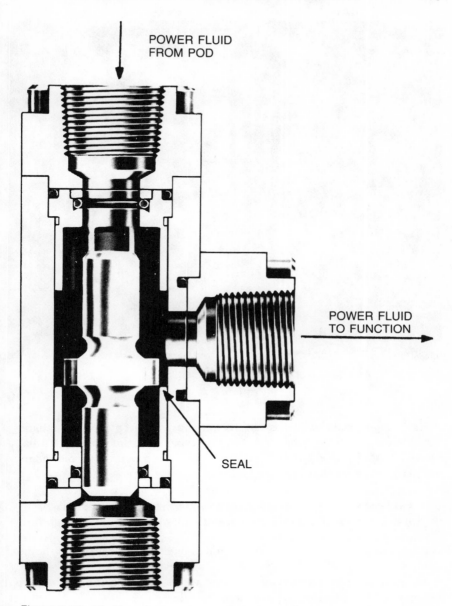

POWER FLUID
FROM POD

POWER FLUID
TO FUNCTION

SEAL

Figure 5-19. Shuttle valve.

Figure 5-20. Hydraulic unit for subsea BOP control.

Let us go back to the meaning and basic concepts of redundancy. Redundancy is defined by Webster as having more of an item (usually words) than is necessary. However, in offshore operations, redundancy should be defined as:

> *Two nearly identical components combined to perform a specific operation even if either component fails.*

Actually, just having two-of-a-kind is not enough: It is the way that they are combined that is important.

For simplicity, consider two two-way valves. When arranged in series, either valve will shut off flow even if the other valve fails to open. But if either valve fails closed, flow cannot be resumed until both valves have been opened. When these valves are arranged in parallel, either valve will open the line even if one valve has failed closed. However, if one valve fails open, the line will remain open until both valves are closed. The concept of series and parallel redundancy is basic to redundant hydraulics. A combination of these simple basics when applied to more sophisticated series of valves is

PRINCIPLE OF REDUNDANCY

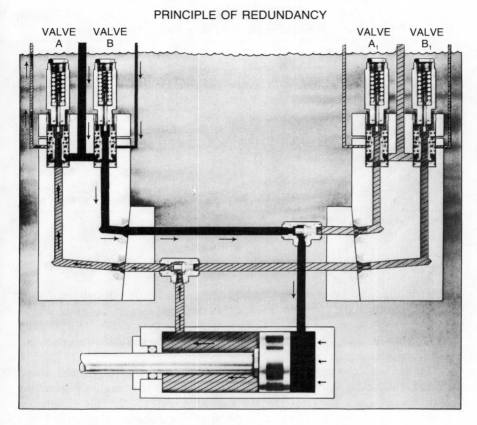

Figure 5-21. Example of redundancy between pods on a stack.

used to obtain the high degree of redundancy present in BOP control systems.

The parallel system is represented by two separate pods on the stack. Separate lines are run to each pod from the surface. Control signals are sent to the active pod only. The valve or switch between the two parallel systems cannot be redundant; therefore, it is located at the surface so that it can be replaced with relative ease. To see how this redundancy works on the stack, we will analyze the operation of a simple, single function, to see how it is operated by each pod. This analysis may seem tedious, but understanding this part of the operation is important to understanding the overall system and diagnosing problems that may occur.

The system to be discussed is shown in Figure 5-21. In this figure are two pods, which we shall call the L (left) and R (right). We shall call the movement of the piston to the left *close* and to the right *open*. Valves B and

Table 5-2
Fluid Directions Between Pods in a Redundant System

Seq.	Active Pod	Open/ Close	Power Valve	Vent Valve	Shuttle Valve Operation *	
					Open	Close
1	L	Close	B	A OR B_1	No shift	No shift
2	L	Open	A	B	No shift	No shift
3	R	Close	A_1	A	No shift	Shift
4	R	Open	B_1	A_1	Shift	No shift
5	R	Close	A_1	B_1	No shift	No shift
6	L	Open	A	A_1	Shift	No shift

*Shuttle valves are shifted by power fluid only, not by venting.

A_1 are connected through a shuttle valve to the *close* side and valves A and B_1 are connected to the *open* side.

With pod L active, opening valve B directs power fluid through the shuttle valve and closes the function. Fluid from the open side moves through the shuttle valve and, depending on the position of the shuttle, is vented through valve A or B_1. Table 5-2 shows the operation of the shuttle valves during a specific sequence.

On the surface, redundancy is similar between control panels. Hand operated valves or solenoid valves allow air pressure to operate a piston, just as hydraulic fluid operates the function in Figure 5-22. This air piston, however, operates a rotary valve on the hydraulic unit (see "Hydraulic Unit" earlier in this chapter) that pressurizes a pilot line with hydraulic fluid, operating a specific pilot valve on the subsea pod. The operations of closing and opening a ram are shown in Figures 5-22 and 5-23. *Block* position is a third position, used primarily on direct operating *surface* BOPs to produce a hydraulic block in a line. With the more sophisticated indirect and remotely operated *subsea* BOPs, the *block* position is not well defined. *Block* operation varies between floaters and should be used with care if at all. Operators do not generally use this position because of uncertainties that may arise.

Operating Pressures and Capacities

The operating pressure of the hydraulic unit is about 3,000 psig. The subsea accumulators also operate at the same pressure above ambient sea pressure. This pressure is used because it has been the highest pressure for standard, off-the-shelf equipment. The greater the pressure differential, the

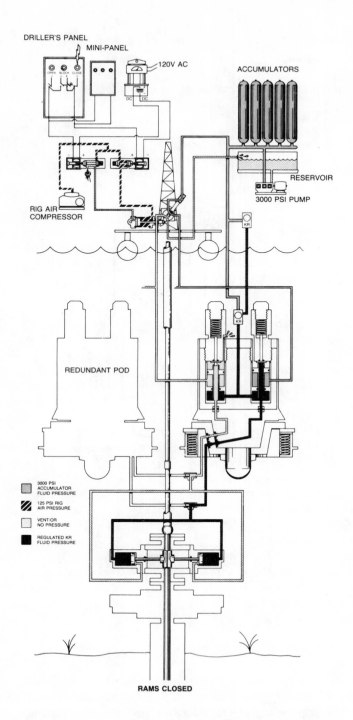

Figure 5-22. Flow paths to close rams from the driller's panel.

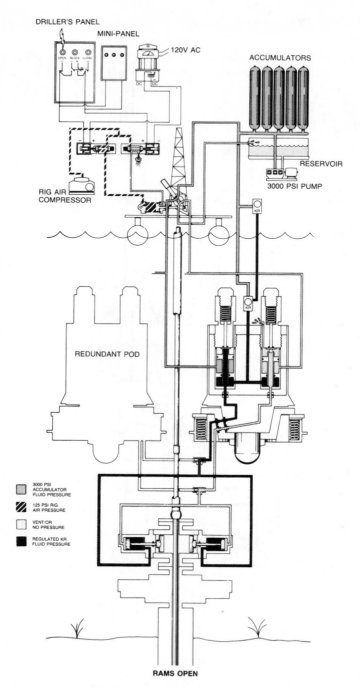

Figure 5-23. Flow paths to open rams from the driller's panel.

higher the fluid flow and the faster the reaction times. The stack functions, however, are designed to operate at a lower pressure.

The industry-accepted operating pressure for the majority of the functions on the stack is 1,500 psi above ambient pressure. For modern equipment this includes:

Shear rams (3,000 psi was required on the early rams)
Pipe rams
K&C valves (lower pressures on some of the early valves)
Hydraulic connectors

A hydraulic regulator controls this pressure.

When stripping tool joints through annular preventers, operating pressures less than 1,500 psi are needed to prevent damage to the rubber element. The closing pressure for stripping depends on the preventer design and size, the well pressure and the size of the pipe being stripped. Thus, a special regulator for the annular preventer(s) should be included with each pod. These regulators must have good flow characteristics, because annular preventers require large closing volumes. The regulators should not be expected to handle the high rate, low volume flows encountered when stripping a tool joint through an annular preventer.

Accumulator Capacity Requirements

Adequate accumulator capacity is essential for reliable operation of a BOP stack, but it is not the only point of consideration (see "Pumping Capacity" section below). The criterion recommended by API* is an accumulator volume 50% in excess of the fluid requirements to open and close *all* functions on the stack. This usually includes all rams, one annular preventer and the K&C valves.

Accumulator gas precharge (gas pressure with the liquid side open to the atmosphere) of 1,000 psig is generally accepted, because it is at least 40% higher than the minimum pressure needed to operate any function on the stack including the pilot valves. Fluid storage pressure is conventionally 3,000 psig maximum. It is the maximum pressure rating for off-the-shelf equipment, and is an economic upper limit. These economics may change in the future for deepwater drilling where the efficiency of the 3,000 psi system decreases.

The example worksheet (Table 5-3) shows how to calculate fluid volume requirements for surface accumulators. The fluid capacity of an accumulator

---

*This API recommendation is not accepted by all governments.

Table 5-3
API Recommendations
Fluid Volume Requirements for Surface Accumulators

Nominal stack size 16¾, pressure rating 5,000 psi, K&C Valve size 4 inches

Function	Make	No. of Units	Volume per Function (gal)	Volume Requirements (gal)
Rams close	Cameron	4	10.6	42.4
Rams open	Cameron	4	9.8	39.2
Annular close	Hydril*	1+	51.1	51.1
Annular open	Hydril*	1+	33.8	33.8
K&C valves open	McEvoy	6	1.3	7.8
K&C valves close	McEvoy	3‡	0.9	2.7

Liquid volume to open and close functions	177.0 gal
Safety factor	X 1.5
Liquid volume required	265.5 gal

Precharge (P_c) = 1,015 psia
Working pressure (P) = 3,015 psia

Volume factor = $1 - \dfrac{P_c}{P}$ = 0.663   divide by 0.663

Total fluid capacity required	400 gal

*Hydril GL with secondary port connected to closing line, surge tanks not included.
+Assumes that only one annular preventer will be closed.
‡Fail close option used on three K&C valves.

is defined as the total fluid (liquid and gas) that the accumulator will hold. It must exclude the volume of the bladder (or piston) and the volume of internal valves and other internal objects.

Usable fluid is the amount of pressurized liquid that an accumulator can hold. It depends on the gas precharge as well as the fluid capacity. The relationship is defined as:

$$\text{usable fluid} = \text{liquid capacity} \; (1 - P_c/P)$$

where:

P_c = precharge pressure, abs units
P = working pressure of the system, abs units

The volume factor $(1 - P_c/P)$ shows how higher working pressures increase the volumetric efficiency.

Technically, the pressures for this equation should be in absolute pressure, but at the pressure levels used, ambient pressures may be neglected at the surface. Subsea ambient pressures, however, are appreciable.

Subsea accumulators are mounted on the stack. Often some accumulators will be dedicated to emergency operation of the stack in the event of a failure of the control lines to the surface. Additional accumulators are used to improve reaction times by decreasing the length of line between the fluid source and the functions.

Precharge pressure should be 1,000 psi above absolute ambient pressure for subsea accumulators or 1,015 psia (at the surface) + 0.5 × water depth (in feet). Now,

$$\text{volume factor subsea} = 1 - \frac{1,015 + 0.45 \text{ water depth, ft}}{3.015 + 0.45 \text{ water depth, ft}}$$

for a 3,000 psig system. Figure 5-24 shows the variation in volume factor with water depth. It is evident from this figure that the usable fluid in an accumulator decreases with increasing water depth when using the same operating pressure at the surface.

Pumping Capacity

The philosophy of sizing the pumps for the hydraulic unit varies between operators. API recommends enough hydraulic horsepower to close the annular preventer in 30 seconds. Other criteria are based on the recharge rate of the accumulators. Either criteria usually result in requirements of 60 to 90 horsepower for the newer rigs. Modern rigs use two 30 or 40 hp pumps and include two or three air driven pumps for backup.

With 80 horsepower of pumping capacity, at 80% overall efficiency, a system such as described in Table 5-3 would be able to deliver about 540 gallons of fluid in five minutes, enough to open and close all functions three times. If this much fluid is used, you must have a severe leak that must be corrected. Otherwise, you have an even more severe problem, but not with the equipment.

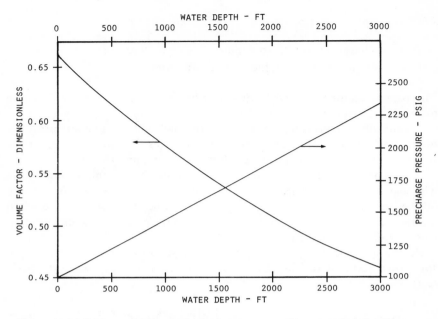

Figure 5-24. Volume factor and precharge pressure vs. water depth.

Factors Influencing Reaction Times

Reaction times are being improved by a combination of brute force and technology. *Pilot hoses* have a major effect on reaction times, but there is no conventional flow in these lines. Fluid is injected into the line, and when the line pressure is high enough, the pilot valve is actuated. Pressure is bled off later to close the valve.

Reaction times are increased by the expansion of the pilot lines. Generally, production processes have allowed a better reinforcement of the 5/16-inch lines than the 1/4-inch lines. The best way to determine the expansion is experimentally. Usually, manufacturers of the control system will have line expansion data available along with the minimum pressure required to actuate the pilot valves. The most current data should be used for the most realistic evaluation.

Pilot lines should be considered as sumps and cleaned out often. Precipitate, pipe dope, Teflon tape, metal filings and heaven knows what else have been blown from the pilot lines when cleaning. Some hydraulic fluids will precipitate when mixed with water of high mineral content. This precipitate will naturally collect at the bottom of the line. New control systems should be checked the first two or three times the stack is on the surface.

Filters are used in the hydraulic system to collect most of the unwanted material. These filters are very important but do not eliminate the need for cleaning the pilot lines. Filters should be changed according to the control system manufacturer's recommendations.

Pressure losses caused by friction in the plumbing is a major source of decreased reaction times. Fluid flow is difficult to define and various books[1,2] have written on the subject. A simple approach can be taken by considering the friction losses as described by the Fanning or Darcy equation[3] that may be expressed as:

$$F = K \frac{f \, L \, q^2}{D^5}$$

where:

F = friction loss in the line
L = line length
q = flow rate
D = line diameter
K = constant for unit conversion
f = friction factor, function of Re

Re = Reynold's number = $\dfrac{Dv\rho}{\mu}$

v = fluid velocity
ρ = fluid density
μ = fluid viscosity

This defines why some of the factors, most of which have already been mentioned several times, are so important. Line lengths should be kept as short as possible. The biggest problem with line lengths seems to be in the plumbing on the stack. Increasing the line diameter appreciably decreases the resistance to flow, so the largest reasonable pipe size should be used for the power fluid. We can do nothing about friction loss with flow rate, because we want the highest rate possible. The friction factor[4], f, depends on the flow regime. In the range of flow rates of interest to BOP control systems, f decreases with increasing Reynold's number. This is an indication to keep the viscosity as low as possible.

An *ell* in the line has the effect of lengthening the line by about 32 pipe diameters[5], the equivalent of 2.7 feet in a 1 inch ID line. A *tee* is the equivalent of adding 60 pipe diameters or 5 feet to a 1 inch line. This assumes that the friction loss results from a change in direction of flow, and that there are no restrictions in the fittings.

The *difference in differential pressure* between working and ambient pressure is essentially the same at the surface and subsea. The density of the control fluid is about 63 lb/ft³ (1.01 SG) compared with sea water at about 64 lb/ft³ (1.026 SG), a pressure increase of about 1.6% or about 21 psi (1.5 kg/sq cm) in 3,000 ft (909 m) of water.

BOP Testing

The BOPs are the primary means of controlling the well in the event of a kick. The importance of their proper operation cannot be over-stressed. All personnel aboard should be interested in the stack's operation: Their lives may depend on it.

The stack will be extensively tested at the surface before running, and periodically tested during breaks in the drilling operation. The frequency of subsea testing may vary between operators. Local laws usually specify a minimum frequency for testing.

Surface Testing

Testing and maintenance at the surface must be thorough, because the stack will be required to operate subsea for several months. A *test stump* that is the male profile of the wellhead is used for testing at the surface. The hydraulic connector at the bottom of the stack is latched onto the stump, and all seals on the stack—including the AX-ring—are checked for leaks. The machined portion of the stump should be protected by a cover when not in use.

While testing the BOPs at the surface, observe the following rules:

1. Procedures should include special diagnostic tests recommended by the manufacturers of each component on the stack.
2. The control system should be pressure tested before the BOPs.
3. Tests should be made with water, not mud, in the BOPs.
4. All components should be tested at the rated working pressure.
5. All pipe rams and annular preventers should be tested by closing around drillpipe to avoid damaging the seals.
6. Secondary seals on the rams must be inactive during the pressure test. The primary seals cannot be tested if the secondary seals are active (see "Ram," earlier in this chapter).
7. All ram locks should be tested by locking the rams and bleeding off the hydraulic closing pressure while holding rated BOP pressure below the rams.
8. All hydraulic connections and stack connectors should be carefully inspected for leaks.

9. It is advisable to keep a log of reaction times for the major stack functions. Changes in reaction times denote changes in the hydraulic efficiency of the control system. These changes may indicate impending failure.

Subsea Testing

Subsea testing of the BOPs cannot be as extensive as surface testing. A test plug (see the section on "Casing Seal Pressure Tests" in Chapter 3) is used to seal off the casing while pressure testing. The plug is run on drill-pipe. Flow from the pipe indicates a leak in the test plug. The shear/blind rams are function tested at the specified periods, but no method is available for pressure testing the shear/blind rams without running a high risk of rupturing the casing.

Pressure Integrity of the BOP Stack

Originally, BOP stacks were designed for land use. On land, the forces of major importance were the well pressure acting on the sealed off area, the weight of the stack above each clamp, and the resistance of this clamp to retain the forces caused by well pressure.

With the advent of subsea wellheads that require drilling risers, additional horizontal and vertical forces were imposed on the top of the stack. Hydrostatic mud pressure caused by the mud in the riser is also significant, as is the hydrostatic seawater pressure. The results of these forces tend to decrease the pressure rating of the stack.

The effect of external forces on the pressure rating of a stack was described by Bednar, Dixon and DuMay[6] in terms of the measurable parameters.

If we have a stack (see Figure 5-25) with a clamp below a closed ram (see Figure 5-26), the forces will act on this clamp as shown in Figure 5-27. The pressure that this clamp will retain will be described as:

$$P = \frac{1}{\pi R_G^2 + C_R} \left(-\cos\theta - \frac{2L}{R_N} \sin\theta \right) T_B \qquad \text{\textit{tension and bending loads}}$$

$$+ \frac{1}{\pi R_G^2 + C_R} \left[F_C - C_R (P_M - P_W) \right] \qquad \text{\textit{pressure and end load retaining forces}}$$

$$+ \frac{1}{\pi R_G^2 + C_R} \left[W + (P_W A_W - P_M A_M) \right] \qquad \text{\textit{stack weight and hydrostatic pressure forces}}$$

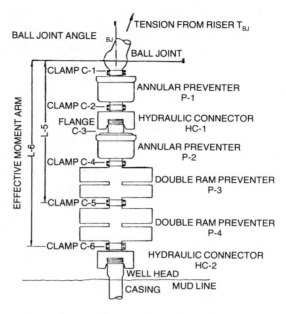

Figure 5-25. Schematic of the BOP stack.

Where:

A_M = projected area where drilling mud acts, in.^{2}
A_W = projected area where sea water acts, in.^{2}
C_R = ring gasket retaining coefficient, lb/psi
F_C = total clamping preload on hub faces, lb
L = vertical distance from horizontal centerline of the ball joint to the bottom connector, ft
P_G = pressure differential across ring gasket, psi
P_M = hydrostatic pressure from drilling mud, psi
P_W = hydrostatic pressure from sea water, psi
R_G = ring gasket sealing radius, in.
R_N = nominal bearing radius between clamp and hub lips, in.
    = tension at ball joint = applied tension - in water weight of riser, ln
W = weight of stack that is above the lowest clamp, lb
O = ball joint angle, deg

This equation can be applied to any of the connectors on the stack, but the moments caused by horizontal forces make the bottom clamp the

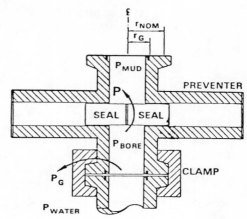

*Figure 5-26. Pressure differentials
acting on preventer.*

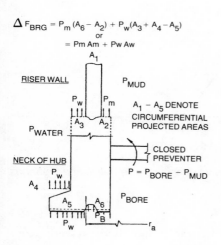

$$\Delta F_{BRG} = P_m (A_6 - A_2) + P_w(A_3 + A_4 - A_5)$$
$$\text{or}$$
$$= Pm\ Am + Pw\ Aw$$

*Figure 5-27. Effect of hydrostatic
pressure on hub bearing load.*

limiting factor. This calculation is for clamps that are properly assembled
and torqued to specifications.

Results of these calculations are shown in Figure 5-28[6] for a COP stack in
3,000 feet of water. The results vary little with water depth.

Backup for BOP Control

In floating drilling operations, it is advisable to be able to close the BOPs
if the primary systems fail. Originally, divers accomplished this task, but for
deeper water, an automatic closing system was developed. The system now

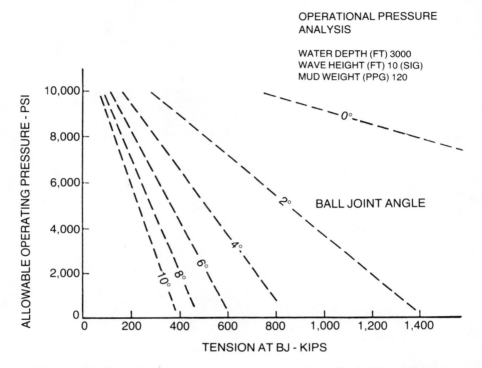

ALLOWABLE OPERATING PRESSURE
CONSTANT BALL JOINT ANGLES

OPERATIONAL PRESSURE
ANALYSIS

WATER DEPTH (FT) 3000
WAVE HEIGHT (FT) 10 (SIG)
MUD WEIGHT (PPG) 120

Figure 5-28. Allowable operating pressure vs. tension at the ball joint, for various ball joint angles.

in use closes an annular preventer when the power fluid pressure drops below a predetermined level.

In the 1970s, acoustic backup systems were developed by TRW, Raytheon, and the Payne division of Cameron Iron Works. These systems have the advantage of being able to remotely operate selected functions on the stack.[7] Power fluid is furnished by dedicated accumulators located on the stack. Part of the fluid is diverted through electric solenoid valves to operate pilot valves. Electrical energy is furnished by batteries stored on the stack. These systems are similar to the multiplexed electrohydraulic system, but they receive signals acoustically instead of through electric lines. A typical surface command unit is shown in Figure 5-29. Most surface command units are portable.

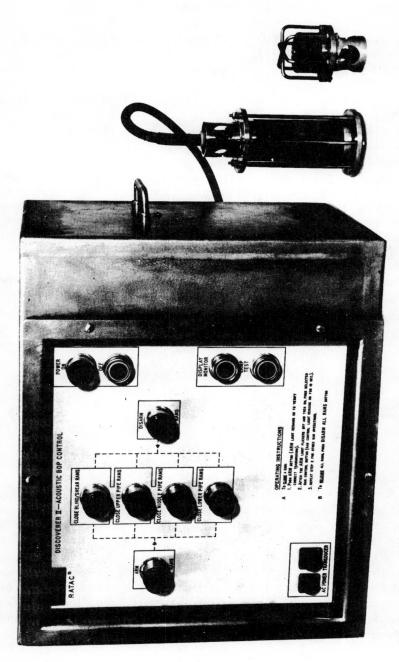

Figure 5-29. Acoustic backup for BOP control—surface unit.

References

1. Prandtl, L., and Tietjens, O.G., *Fundamentals of Hydro & Aerodynamics*. Dover Publications, New York, 1957.

2. Byrd, B.R., Stewart, W.E., and Lightfoot, E.N., *Transport Phenomena*. John Wiley & Sons, New York, 1960.

3. Perry, J.H., *Chemical Engineers' Handbook*. McGraw-Hill Book Co., New York, 1950. p. 377.

4. Badger, W.L., and Banchero, J.T., *Introduction to Chemical Engineering*. McGraw-Hill Book Co., New York, 1955. p. 27.

5. Perry, *Chemical Engineers*, p. 390.

6. Bednar, J.M., Dixon, W.P., and DuMay, W.H., "Effects of Blowout Preventer End Connections on the Pressure Integrity of a Subsea BOP Stack Under Riser Loads," OTC paper 2649. Presented at the Offshore Technology Conference, Houston, 1976.

7. Sheffield, J.R., and Wienzek, W.F., "Using Noise Data from Underwater Gas Vents to Estimate the Noise Level of Subsea Blowouts," *Jour. of Petr. Tech.* (March 1976). p. 259.

8. Altermann, J., III, "Practical Considerations for Arranging, Testing BOP Stacks, *World Oil*, May 1980, pp. 91-104.

The Drilling Riser

6

The marine riser or drilling riser has probably been "cussed" as much by its users as the original *grief stem* (early slang for the kelly) was; but in present-day floating drilling, the riser is essential. Improvements in materials and construction techniques since the mid 1960s have decreased the probability of riser failure; however, without proper care and use, accelerated riser deterioration can result.

The riser system is the communications link between the vessel and the subsea wellhead. Through this riser, downhole equipment is guided into the well and mud is returned to the surface.

As discussed earlier, there are six degrees of freedom of motion on a drilling vessel. Without these motions, just a riser pipe like those used on platforms would be sufficient. However, because of the motions, additional equipment is required (see Figure 6-1).

The *slip joint* compensates for the heave of the vessel and consists of an *inner barrel* (1) that slides into an *outer barrel* (2). The inner barrel is connected to the vessel by some mechanism—a ball joint or a gimbal—that will allow the vessel to pitch and roll without contorting the riser. The mud flowline and the diverter system will be between the inner slip joint barrel and the rig floor.

The *outer slip joint barrel* supports the entire riser. Riser tension is maintained through tensioners that are connected by wire rope to a tensioning ring that butts up against a hub on the top of the outer slip joint barrel. The flexibility of the wire rope minimizes the effects of yaw motion that would otherwise be transmitted to the riser.

Sometimes in deepwater drilling, an *upper ball joint* (3) is used. This ball joint is used to decrease the stresses caused by riser motion acting on the transition of stiffness that occurs when going from the riser to the slip joint. Use of the upper ball joint is now being questioned: Wells have been drilled in over 1,500 feet of water without this additional ball joint.

The main body of the riser is comprised of *riser joints* (4), or *stalks*, with *kill* and *choke lines* (5) integrally attached.

The *lower ball joint* (6), referred to as the *ball joint*, is located between the bottom riser joint and the BOPs. It allows for limited deflection of the riser and compensates for the surge and sway (offset) of the vessel.

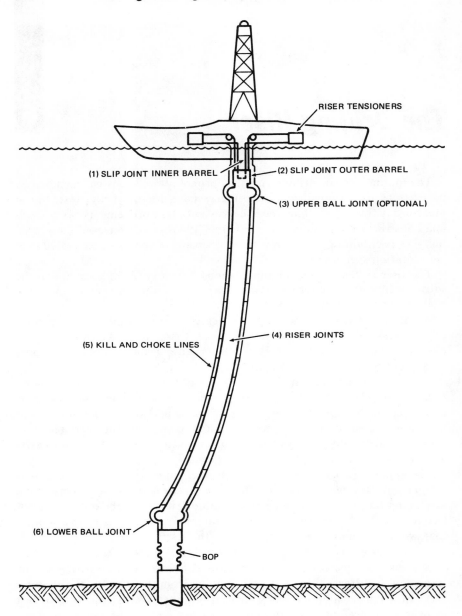

RISER TENSIONERS

(1) SLIP JOINT INNER BARREL

(2) SLIP JOINT OUTER BARREL

(3) UPPER BALL JOINT (OPTIONAL)

(5) KILL AND CHOKE LINES

(4) RISER JOINTS

(6) LOWER BALL JOINT

BOP

Figure 6-1. Riser system for a floating drilling rig.

Figure 6-2. A riser joint.

Riser Components

Components of a riser system must be strong enough to withstand high tension and bending moments, and have enough flexibility to resist fatigue, yet be as light as practicable to minimize tensioning and floatation requirements. These considerations should be given to all components when selecting a riser system. Major manufacturers of riser components are Regan International (Hughes), Vetco Offshore, National Supply and Cameron Iron Works. NL Rig Equipment (Shaffer) has a riser system designed but, it is not available at this printing.

Riser Joints

A *riser joint* is constructed of seamless pipe with mechanical connectors welded on the ends. Kill/choke lines are attached to the riser by extended flanges of the connector (see Figure 6-2). Sometimes, lines for BOP power fluid are integrated, as are the kill/choke lines. The riser can be run in a manner similar to drillpipe: by stabbing one stalk at a time into the string and tightening the connector.

The female portion of the connector contains dogs or ring segments that are actuated by bolts, and clamp into grooves machined into the male portion of the connector. Two to six bolts (depending on the manufacturer) are used for each connector. Increasing the number of bolts increases the time for making a connection. This is one—but certainly not the only—factor to be considered when running long strings of riser.

Welding and heat treating the connectors and pipe when the riser joints are manufactured is very important in maintaining the fatigue properties needed for riser integrity.[1,2] No field welding should be permitted on a riser.

Ball Joints

Ball joints or flexible joints allow limited angular motion of the riser. In some cases, these flexible joints may be a series of ball joints. *Pressure compensated* ball joints should be used to decrease torque required to deflect the joint. Forces acting on the joint push the inner ball against the outer casing, causing the joint to bind (see Figure 6-3). To decrease the required torque, hydraulic fluid is injected to spread apart and lubricate the moving parts. With the large areas involved, relatively small pressures are required. In fact, pressure regulation is critical because over-pressuring will bind the joint just as underpressuring will. The compensating pressure is a function of mud weight, riser tension, and the internal areas on which the mud pressure acts.

In the late 70s, two new ball joints were designed and are in use on some vessels commissioned after 1976. The Cameron Universal Ball Joints are designed in two models and will withstand 1,000,000 pounds of tensile load. The Model "DW" and the Model "SW" are designed for 6,000 feet and 2,000 feet of water, respectively.

The load is carried by large U-joint pins, and the seal is a sliding resilient seal. The Vetco Uniflex Joint is designed for 1,500,000 pounds tensile load and for 10,000 feet of water. The moving parts are a series of elastic and steel plates bonded together to both seal and allow for joint flexure.

Since the drillpipe and other tools are allowed or forced to bend through the ball joint, wear bushings are used. One bushing is always inside the wellhead during drilling (see Chapter 4). An optional upper bushing is used by some contractors to protect the riser near the ball joint. Protective bushings must be retrieved before casing is run.

Slip Joints

A *slip joint* is comprised of two concentric cylinders or barrels that telescope (see Figure 6-4). The outer barrel is attached to the marine riser, and the riser is held in tension by wire ropes from the outer barrel to the tensioners. On most vessels, the tension lines are connected to a ring that bears against a flange at the top of the outer barrel. The ring is held in place

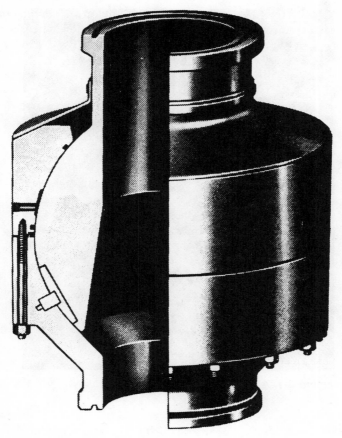

Figure 6-3. Pressure compensated ball joint.

when the riser is tensioned, but will slide along the slip joint when tension is relieved. This saves time when installing the slip joint. For dynamically positioned vessels, a bearing surface may be used between the tensioner ring and the flange to allow the ring to rotate at low torque.

The inner barrel is polished steel and is attached to the diverter. If a diverter is not used, the inner barrel may be attached with wire rope slings beneath the rig floor.

Rubber packing seals the annulus between the inner and outer barrels to hold the mud in the joint. The packing is actuated hydraulically or with air pressure. In practice, a very small mud leak should be allowed to lubricate the packing. Slip joints should be locked shut when handled on the deck.

Figure 6-4. A slip joint.

Diverters

A *diverter* should be installed between the inner slip joint barrel and the vessel. A diverter is a low pressure annular preventer that seals off the riser bore (see Figure 6-5). This diverter redirects the flow of mud and cuttings that would otherwise be blown on the rig floor during a kick when the BOPs are not used. The diverted fluids flow overboard. This system should be designed to withstand high velocity impact of sand and cuttings, not high pressure. Flow lines from the diverter should have large IDs (10 inches or larger) to minimize back pressure at high rates of gas and mud flow. Diverter systems are customarily designed for 50 or 100 MMscf/d gas flow.

The control system for the diverter should be designed to open the flow lines automatically when the diverter is closed. *The riser should not be shut-in, because relatively low pressure in the riser can cause buckling.* Two lines should be available so that discharged fluids can be diverted through a leeward line.

Figure 6-5. A diverter system.

Design considerations for mechanical strength of the diverter and gim-balling mechanism include the upward force from the back pressure caused by flow and particle impact on the diverter area and the inner slip joint area.

Jumper (Transition) Lines

Jumper lines are the transition of the high pressure kill/choke lines around the ball joint and slip joint. For 5,000 psi stacks, rotary hoses are used. For stacks with higher pressure ratings, flex-joint (chicksan) lines may be used. Recently, hose manufacturers have developed hoses that are rated to 10,000 psi working pressure.

High flexibility is needed for slip joint jumper lines where large amplitude motions occur. At the ball joint, motion amplitude is relatively low and steel lines are popular as jumpers for high pressure systems. These steel lines may be formed in either vertical steel loops (see Figure 6-6) or in helices (see Figure 6-7). Both vertical loops and helices are in field use.

Buoyancy Modules

Estimated conditions at a specific site may require more tensioning than is available on the drilling vessel. Buoyancy or flotation modules can be attached to the riser to decrease the tension required at the surface.[3] These modules may be thin-walled air cans or fabricated syntactic foam modules that are strapped to the riser. As might be expected, each has its own advantages and disadvantages.

Air cans have a predictable buoyancy, and this buoyancy can be controlled from the surface. However, they slow down the running operation more than syntactic foam modules do. Before running, air cans are installed on the riser joint(s). When run, the cans are interconnected by tubing. Water is allowed to fill the cans so that internal pressure is ambient sea pressure. After the riser has been run, air can be used to displace a controlled amount of water from inside the cans. The number of modules and the amount of water forced from the cans determines the amount and position of the buoyancy. Air cans have been used on deepwater vessels since 1975.

Syntactic foam modules began to appear in the early 1970s. They are convenient because they become part of the joint, they require only a little additional care in handling, and they do not have lines to be attached. Various compositions of foam have been tried with varying degrees of success that are hard to evaluate. Syntactic foam degrades with time and water depth. Determining the degree of degradation requires weighing the samples in a pressurized water chamber—a tedious test that requires expensive equipment. Various modules are in field use, but reliable methods for predicting their buoyancy in operation have not been devised.

Riser Tensioning

The preferred riser tenion can be defined for any given situation as the tension that will minimize the probability of damaging the riser or drilling equipment, yet cause minimal wear to the tensioners. As the definition implies, judgment is required to determine the preferred tension.

Figure 6-6. Vertical steel loops used for kill/choke line transition around the ball joint.

Figure 6-7. Horizontal helices used for kill/choke line transition around the ball joint.

Adequate tension must be held on the riser to insure riser integrity and to keep the riser system (including the ball joint) straight enough to drill without the drillpipe's rubbing and damaging the riser and wellhead equipment. An important property of steel is that it can withstand tension but is easily bent. This can be demonstrated qualitatively using a wire coathanger. You can pull on it very hard before deforming it, but it can be bent with relative ease by simply pushing on the extreme ends. So it is with risers, as will be discussed.

The importance of riser tension has been discussed since the inception of floating drilling.[1,2,4,5] Computer programs are required to model riser stresses as a function of tension. Major oil companies and some drilling contractors

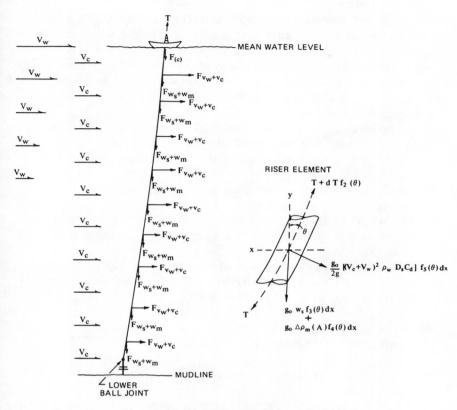

Figure 6-8. Schematic diagram of riser with imposed forces.

have such programs. A field method for determining riser tensioning requirements was developed by Sheffield and Caldwell.[6]

Both static and dynamic riser computer programs are available. The API has compared nine static programs. They have also compared eight dynamic riser computer programs. Comparisons were based on specific models used for the riser. How well the programs agree requires judgment based on the proposed use of the resulting output. Results of the comparison were published by API in 1977.

The marine riser is analogous to a horizontal steel beam that is supported at both ends and loaded between the supports. It will sag, but the sag will be less if we pull on both ends of the beam.[7] A riser is similar, but it is nearly vertical and the loading consists of forces resulting from the waves, current, riser weight and weight of mud in the riser. These forces are shown schematically in Figure 6-8. As the vessel offset increases, the mud density and riser weight become more important. The riser will sag in a manner similar to the horizontal beam, and tensioning likewise will decrease the sagging.

Sag combined with the surge and sway of the vessel will impose extremely high stresses in the riser joint just above the stack if the ball joint is not operating. When the ball joint is *on stop* (fully deflected), it will not operate, and riser failure can result.

Three modes of operation were defined with criteria relative to the station keeping system and vessel offset (see Chapter 3). With respect to the riser, the criteria for defining these same modes of operation are:

1. Operational: When the ball joint angle is less than 4°.
2. Non-operational but connected: When the ball joint angle is greater than 4°. The ball joint must be kept *off stop* as long as the riser is connected, usually until the vessel has reached an offset of 10% of water depth.
3. Disconnected: The riser is disconnected from the wellhead, and concern for riser tension is replaced by concern for tensioning and paying out the guidelines if the vessel is moved.

To describe what happens to the riser when it is tensioned, it is easier to start with the ball joint *on stop* and discuss situation 2 first.

Non-Operational but Connected Mode

Figure 6-9 is a typical example of maximum/minimum stress behavior in a riser when the ball joint is on or near the stop. When *on stop*, maximum and minimum stresses will occur at or near the connection between the riser and ball joint. When the ball joint is *off stop*, maximum stresses may occur at another position in the riser. In Figure 6-9, the ball joint is just *off stop* when the maximum and minimum stresses at the ball joint become coincidental. For this case, the ball joint disengages at about 100 kips. Increasing the tension to 150 or 190 kips has little effect on the riser stress, but *decreasing* the tension to 50 kips could lead to riser failure. The *minimum* tension to be allowed for the conditions stated in the figure is 100 kips.

Riser tension cannot be held constant in floating drilling operations, and the applied tension will fluctuate about a mean setting. This fluctuation is considered by many in the industry to be about 15% of the mean tension. As 100 kips is the minimum allowable tension, then the tension should be set above 100/0.85, or 118 kips. At a setting of 118 kips, the tension is expected to fluctuate between 100 and 136 kips. This is the *calculated* tension setting to minimize the riser stresses for 10% offset and the conditions listed. Lower offsets will require lower tension settings using the same criterion; however,

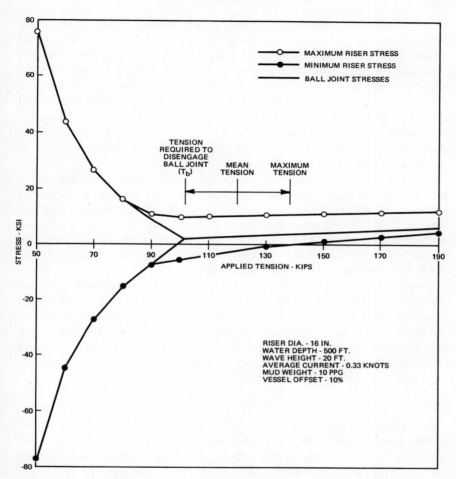

Figure 6-9. Typical curve of stress vs. applied tension of risers—non-operational mode.

tensioners are not designed to change tension rapidly if something should happen. Thus tension settings for field operations should be based on the possibility of a failure of systems that will affect riser stresses (see section on "Operating Factors," later in this chapter).

Operational Mode

Four degrees is the maximum ball joint angle that can be tolerated without damaging the riser and BOPs during drilling operations. The four-degree angle is based on experience combined with experimental data of

Childers and Ilfrey,[8] who established agreement between their findings and those of Lubinski.[5,9] In practice, this operating angle will coincide with vessel offsets of three to four percent of water depth. It is impractical if not impossible in normal field operations to sustain a ball joint angle with the top of the riser offset to six percent of water depth (3.3 degrees offset angle). Thus, to avoid damage to the riser and BOPs, the maximum offset will probably be three to four percent of water depth when drilling or running large tools. Running casing may require less offset and lower ball joint angles than described. This will be for short times only, but such situations *can* be expected in field operations.

For a given riser system on a specific vessel, graphs or tables for determining tensioning requirements under various field conditions should be available. The tensions read are only a starting point and may be low because the tensioning data are no more accurate than the on-site estimate of the current and waves. For example, underestimating the current acting on the lower part of the riser will lead to a low estimate for riser tensioning and this will be reflected in the ball joint angle.

It is evident from this discussion that the ball joint angle gives an indication of the riser stresses. The ball joint angle will indicate that something is wrong, but it alone cannot give a complete analysis of the situation. A procedure for using the ball joint angle in conjunction with a position reference system will be discussed later (see "Use of Instrumentation to Protect the Riser" in this chapter).

Riser Appraisal

Anything that is as expensive and as essential to an operation as the drilling riser demands special evaluation and inspection before use. Riser appraisal requires considerable judgment, and should include calculations of tensioning requirements for various situations, including the failure of support equipment that may effect the riser. The following discussion is intended to give an appreciation for items that may cause trouble. In some cases, trade-offs may be available, while some problem areas may require immediate correction. Operating procedures that include tensioning tables and graphs should be based on this appraisal.

Steel Properties and Fabrication Practices

Even when the steel used in a riser is satisfactory from a stress standpoint, its impact strength and toughness may be inferior. Steel inferior in these respects is subject to brittle failure due to cyclic stresses, which means that there is very little lag time between detectable crack initiation and complete failure. For example, Grade B seamless line pipe and J-55 casing are inferior in this respect to X-52 seamless line pipe, because X-52 has reasonably good properties in the required stress range after it has been normalized at the mill. Currently, some deepwater operators' preference is for 80 ksi yield with good impact strength and toughness.

Welding the connectors to the riser should always be done under highly controlled conditions in a shop. Preheating, welding and normalizing after welding have a significant effect on riser integrity and useful operating life.

No field welding on the riser should be permitted.

Fatigue

In the appraisal of a used riser system, the service history of the riser is an important factor. This is because fatigue effects are cumulative, but cannot be measured by current inspection methods. The first indication of excessive fatigue is the formation of detectable cracks in the material. Depending on the material used, crack initiation may be followed closely by failure, particularly if the proposed operating conditions are severe. Thus, even when stresses are minimized by the correct tension, a previously fatigued riser may fail in subsequent service. Experience has shown that with risers constructed of good materials as discussed in the foregoing, routine riser inspections are usually adequate in avoiding failure (see section on "Inspection and Maintenance," later in this chapter).

Fatigue is an embrittlement of the metal. It occurs first at points where stress is concentrated, such as welds. Fatigue is caused by cyclic loading.

For a given material, fatigue depends on the mean stress level and the stress fluctuations about this mean (see Figure 6-10). It can be seen from this figure that both the tension held on the riser to minimize stress and the tensioning equipment can strongly effect riser service. The example is for a specific metal. Increasing the mean stress and the fluctuations about this mean will decrease the useful life of any riser material, but some metals will withstand it better than others.

Fatigue properties are as important for a riser material as the initial tensile strength.

Operating Factors

The *influence of mooring capability and reliability* is an important operating factor. For the operating case, the vessel offset must be held within the minimum operating offset if the 4 degree maximum ball joint angle is to be maintained. For the nonoperational case, too low tension combined with excessive vessel offset can cause permanent riser deformation, and possibly riser failure. Figure 6-11 presents an example of the stress increase that will accompany increasing vessel offset at constant tension. In this case, the

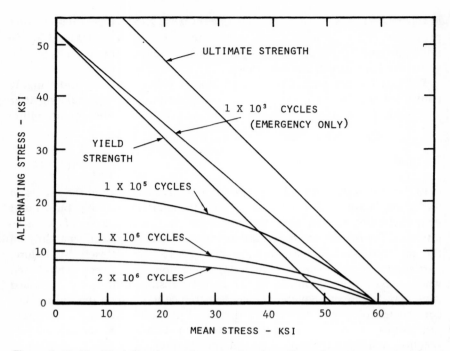

Figure 6-10. Modified Goodman diagram showing effect of mean stress level and stress fluctuations on a specifically treated metal.

maximum riser stress at 225 kips tension will increase from 11 ksi at 6% vessel offset to 34 ksi at 10% vessel offset. In this instance, the riser would be permanently deformed if it were to be constructed with materials having a minimum yield stress of less than 34 ksi. But for this same occurrence, the riser stress level is essentially constant at about 14 ksi for all offsets from 3 to 10% if the recommended tension of 296 kips for 10% offset is used.

These data show that for the non-operational case:

1. Tensioning for the maximum offset minimizes the risk of overstressing the riser.
2. The ability of the mooring to limit vessel excursions under the worst anticipated conditions should be known.
3. Procedures should be devised to monitor vessel position, to manipulate the moorings so that the vessel will be held within the design value and the riser will be released before the design offset value is exceeded.

Tensioning equipment reliability is another factor that should be considered. With multi-unit tensioning systems, it is unlikely that all units will

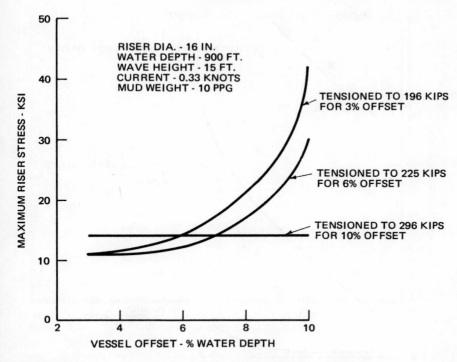

RISER DIA. - 16 IN.
WATER DEPTH - 900 FT.
WAVE HEIGHT - 15 FT.
CURRENT - 0.33 KNOTS
MUD WEIGHT - 10 PPG

TENSIONED TO 196 KIPS
FOR 3% OFFSET

TENSIONED TO 225 KIPS
FOR 6% OFFSET

TENSIONED TO 296 KIPS
FOR 10% OFFSET

MAXIMUM RISER STRESS - KSI

VESSEL OFFSET - % WATER DEPTH

Figure 6-11. An example of the effect of vessel offset on stress at a constant tension.

fail simultaneously. It is possible that two units operating from the same pressure reservoir could fail, but this usually occurs relatively slowly, and with some warning. The breaking of a single tensioning line is more frequent. Figure 6-12 shows the effects of losing one unit of a six-unit system. Obviously, a riser failure can result even with only partial failure of mooring and tensioning systems if these failures occur simultaneously.

A realistic policy of retiring tensioner cables will do much toward increasing tensioner reliability. Particular attention should also be given to sheave sizing and to tensioner mounting design.

Kill/Choke Lines

Kill/choke lines are an integral part of the riser and the flow path for circulating high pressure fluids from the BOPs to the choke manifold. The kill/choke lines should be as large as possible to decrease the back pressure on the formation when circulating out a kick. When a kick is circulated out,

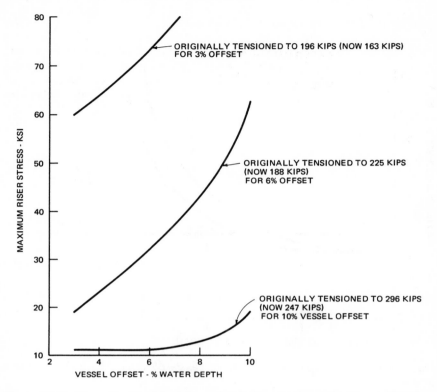

Figure 6-12. The effects of losing one-sixth tensioning capacity on the riser system of Figure 6-11.

the injection pressure is controlled by manipulating the choke. If there is excessive pressure drop in the choke line, it will control the pressure instead of the choke operator. This situation can occur in deep water with the long flow lines required, but how deep is the deep water for this situation? Pressure drop in the lines should be considered when appraising or designing a riser for a given water depth.

At times during operations, it may be advisable to open both lines to decrease the back pressure. This requires some prior planning. The drillpipe has to be positioned so that both lines can be used, and the choke manifold has to be set up with the proper routing for the fluids.

Thus, in operations as well as in design, it is necessary to have a feel for the backpressure to be expected under various conditions. Approximations can be made for typical conditions based on equations that can be used to estimate the pressure drop in the kill/choke lines.

Estimating Pressure Drop in Choke Lines

To determine if the pressure drop in the choke line will be a problem, the power law equation is generally accepted. Moore[10] suggests the following equation to determine if the flow is laminar or turbulent. If the flow is equal to or higher than the critical velocity, F_c, the flow regime is considered to be turbulent (Reynold's number equal to or greater than 3,000).

$$V_c = \left(\frac{5.82 \cdot 10^4 \cdot K}{\rho}\right)^{\frac{1}{2-n}} \left(\frac{1.6}{D} \cdot \frac{3n+1}{4n}\right)^{\frac{n}{2-n}}$$

where:

V_c = critical velocity assuming turbulant flow is fully developed at a Reynold's number of 3,000, ft/sec
K = ordinate intercept or a plot of log shear stress vs log shear rate, lb force/100 ft^2
n = slope ofstraight line plot of log shear stress vs. log shear rate
ρ = mud weight, lb/gal

If readings of mud shear vs. rate are available:

$$n = 3.32 \log \frac{\text{shear reading @ 600 rpm}}{\text{shear reading @ 300 rpm}}$$

$$K = \frac{\text{shear reading @ 300 rpm}}{(511)^n}$$

For *turbulent flow*, the pressure drop may be expressed by the Reed equation as:

$$\frac{\Delta P}{\Delta L} = \frac{7.7 \cdot 10^{-5} \cdot \rho^{0.8} \cdot q^{1.8} \cdot PV^{0.2}}{D^{4.8}}$$

where:

$\frac{\Delta P}{\Delta L}$ = pressure drop per length, psi/ft
q = flow rate, gal/min
PV = mud plastic viscosity, cp
D = pipe ID, in.

This equation assumes that the friction factor in the Fanning equation (see section on "Factors Influencing Closing Times," in Chapter 5).

$$f = \frac{0.046}{Re^{0.2}}$$

$Re = \dfrac{Dv\rho}{\mu}$ = Reynold's number, dimensionless

For equations with very low exponents of Re, $\mu = \dfrac{PV}{3.2}$ may be used.[9]
For *laminar flow*, the following equation may be used:

$$\frac{\Delta P}{\Delta L} = \left(\frac{39.22\ q}{D^3} \cdot \frac{3n + 1}{4n}\right)^n \frac{K}{300D}$$

Straight pipe equivalent for fittings and bends.

For each ell:

$$L_e = 2.7 \text{ pipe ID (in.), answer in ft}$$

For each tee:

$$L_e = 5.0 \text{ pipe ID (in.), answer in ft}$$

For smooth bends,[11] as in jumper lines:

$$L_e = \frac{0.147 \cdot r \cdot \alpha^{1.10}}{D^{1.03}} \left(\frac{q \cdot \rho}{PV}\right)^{0.032}$$

where:

L_e = equivalent straight line length, ft
r = radius of curvature of the centerline, ft
α = angle subtended by the bend, deg

The equation for smooth bends assumes that the bend radius will be greater than six pipe diameters and that the friction factor is the same function as described in the turbulent flow equation.

Inspection and Maintenance

Each riser component, especially riser joints, should receive periodic inspection and maintenance. This requires tagging and keeping records of each riser component. Visual inspections should be made each time the riser is run. Resilient seals should be inspected and replaced when necessary. Each entire joint should be inspected, particularly at the sealing areas. Damaged joints should be sent ashore for additional inspection and repair.

Annual inspections should be performed on all riser components, especially the riser joints. It is preferable, but not essential, to inspect the joints on land, under more controlled conditions than are available on a drilling vessel. Prior to inspection, the joint must be cleaned thoroughly, and it is advisable to sand blast the joints at the welds to remove the paint. All seals should be removed, and the joint visually inspected.

Welds should be inspected using a combination of the following techniques:

Dyes will detect only cracks that penetrate the surface. These are low viscosity oil-base dyes that thoroughly wet the metal and penetrate any cracks on the surface. To use dyes, the paint must be removed from the joints.

Magnetic particle inspection will detect cracks at or near the surface. Magnetic flux density increases at discontinuities in a steel medium. The riser is magnetized, and magnetic particles collect at the cracks. It is advisable to sand blast the areas around the welds prior to magnetic particle inspection.

Ultra-sonic inspection may detect cracks below the surface but may miss cracks that are isolated near the surface. Acoustic signals are very sensitive to density changes in a transmitting medium and are reflected readily by high density discontinuities such as steel and air or steel and water. The transmitter and receiver are both run inside of the riser; thus paint, flotation material or air cans do not have to be removed for inspection.

X-ray inspections are for internal cracks and may miss cracks at the surface. X-ray techniques depend on radiation absorption and must be used on the bare riser only.

No single inspection technique will find all of the cracks that may originate; however, dyes and magnetic particle inspections are the most popular. The decrease in catastrophic failures of risers can be attributed in part to improved riser inspection techniques.

After the welds have been thoroughly inspected, the riser should be repaired and the seals replaced.

Proper handling techniques for tubular goods apply to risers. The connector pin should be protected when not in use. Only lift subs that are designed for the riser should be used.

Hard banding on drillpipe will cause unnecessary wear to the riser and BOPs, and should not be used on a drilling vessel.

Use of Instrumentation to Protect the Riser

A minimum of three instruments are needed to protect a riser on a floating drilling platform. Additional instrumentation is advisable in deepwater drilling, and recommendations for riser instrumentation are being studied by the API.

The minimum instrumentation recommended include:

1. Heave gauge
2. Riser (ball joint) angle indicator
3. *Accurate* method for determining vessel position.

A *heave gauge* is very simple. It is universally accepted by the rig crew, because they usually design and build it, and understand how and why it works. The gauge is a pointer that may be attached to a guideline, and the heave is measured on a graduated board that is attached to the vessel. This gauge is important during rough weather because it indicates when the heave is severe enough to disconnect the riser from the stack. It is also important in determining the proper space-out of the slip joint in areas where tide amplitudes are appreciable.

The *riser angle indicator* sensor is attached to the riser and indicates the riser angle relative to the vertical. Sensor information is transmitted to the surface through wire or acoustically. Usually only one indicator is used. This should be located near the ball joint to measure the ball joint angle.

In the sensor are two pendula that measure the riser angle on two axes: x and y. These axes are fixed with respect to the riser and do not change with vessel orientation unless special modification of the equipment is made.

The angle of interest is the resultant angle of the riser that is calculated from the two angles. When a cathode ray tube is used, θ_x and θ_y angles are read, and the resultant angle must be calculated. With the small angles that are considered, the magnitude of the riser angle can be closely approximated by:

$$\theta \simeq (\theta_x^2 + \theta_y^2)^{1/2}$$

where:

θ = resultant angle
θ_x = riser angle in the x direction
θ_y = riser angle in the y direction

The exact equation is:

$$\theta = \arctan\, (\tan^2\theta_x + \tan^2\theta_y)^{1/2}$$

The error of approximation is only 0.08% at 4° angle, increasing to about ½% at 10°. This is fortunate, because the approximation is what will be read on the cathode ray tube of a commercial riser angle indicator.

The riser angle indicator should be properly maintained and inspected by an electronic technician. The frequency of inspection depends upon the equipment. More frequent, routine calibrations are required to maintain instrument accuracy. This can be performed by a trained member of the rig crew.

Proper orientation of the sensing unit on the riser is important. Tilt measurement is of prime importance, so the unit must be parallel with the center line of the riser. Errors in rotational alignment can be decreased by marking the riser joints and the angle indicator sensor on land and under controlled conditions, with tick marks.

Acoustic position determining systems are the commercially available systems that have the theoretical potential to be accurate enough for floating drilling operations. Their operation was discussed briefly in Chapter 3. On dynamically positioned vessels, these systems are carefully installed, meticulously maintained, and frequently calibrated by an electronic technician. Such luxuries are not available on a moored vessel: This makes the position determining system a weak link in the essential chain of instrumentation for riser protection. Knowing the position of the vessel relative to the well is important for maintaining the riser. Manufacturer's recommendations should be closely followed to maintain the acoustic system in proper working order. Inform the manufacturer of any recurring problems.

An Operating Procedure

A ball joint angle greater than 4° is an indication that something is wrong. The vessel offset may be higher than anticipated or the tension may be too low. A frequent reason for a low estimated tension is that the current and wave forces have been underestimated. In any case, the ball joint angle should be decreased before operations are commenced or continued; otherwise, unnecessary wear to the equipment will result.

If the ball joint angle is above 4° for an operating situation, use the following procedure:

1. Check the position. If the vessel offset is more than 4% of water depth, the offset should be reduced. Use of the riser angle to move a moored vessel is *not* recommended in water deeper than 250 feet.

2. If the ball joint angle is still above 4° with the vessel offset less than 4%, increase the riser tension by about 10% of the riser weight in water. The ball joint angle should decrease by more than one-half degree.
3. Continue to increase the tension in increments of 5 to 10% of the riser weight in water until the resulting change in angle is less than ½°.

Warning: All instruments, including tensioner gauges, must be operating properly. The ball joint angle will help with increasing the tension to minimize riser stresses, but it does not define what these stress levels are. The ball joint angle should not be used to minimize riser tension, because a relatively small change in position of the vessel from the position at which the tension was set could require a higher tension setting (see section on "Operating Factors," earlier in this chapter).

Guidance Systems

There are two basic methods used in floating drilling to locate the wellhead and to guide heavy equipment to the wellhead on the ocean floor. The first method uses wire ropes as *guidelines* to maintain mechanical compmunication between the vessel and the wellhead, and to guide the equipment into position for landing or entry into the well.

Reentry is the term for the second method. It commences without mechanical communications between the wellhead and vessel, using mechanical guidance and alignment only in the final stages of landing the equipment. The reentry technique replaces guidelines on some of the dynamically positioned vessels currently in operation.

Guidelines

The use of guidelines is the best known and most widely employed method for guiding heavy equipment from the drilling vessel to a wellhead on the ocean floor. Unfortunately, simplicity often leads to neglect that can cause expensive delays when the lines are not properly tensioned and maintained. Guidelines have been used in up to 3,461 feet of water, and will provide guidance in deeper water, even when rough seas and strong currents exist.

Guideline selection. As with any state of the art, selection of guidelines is based primarily on experience. The best guidelines seem to be galvanized. 6 X 25 IWRC (independent wire rope core). This is a galvanized version of the standard hoisting rope used on elevators and mine lifts. It is designed to reeve through relatively small diameter sheaves with minimum wear and minimum loss of strength. Table 6-1 shows wire diameters that are being

Table 6-1
Conventionally Used Guidelines
Standard Hoisting Rope, Improved Plow Steel,
6 X 25 Filler Wire, IWRC, Galvanized

Rope Diameter (in.)	Conditions	Water Depth (ft)	Breaking* Strength (KIP)	Tensioning Capacity per Line (KIP)	Weight of 100 ft of Line in Seawater (lb/100 ft)
⅝			27	7	60
	Mild	0–600			
¾			39	7–14	86
	Rough	0–600			
	Mild	600–1500			
⅞			51	14–16	116
	Rough	600–1500			
	Mild	1500+			
1			66	16–22	152
	Rough	1500+			
1¼			103	22+	237
1¾	Discoverer 534	3641	197	60	465

*Data from USS Wire Rope Engineering Handbook, U.S. Steel, Pittsburgh. Corrections from tables for independent wire rope core (IWRC), galvanizing and assuming 4% loss in strength per sheave for five sheaves.

used for different weather conditions and water depths and suggested tensioner capacity. Average breaking strengths are included, and a safety factor of about three is used to determine tensioning capacities.

Care of the Guidelines. The most frequent reason that guidelines break has been inadequate tension. Guidelines are made to take tension, so do not be afraid to pull on them. When running or pulling the riser with or without the stack, the simplest rule is to keep maximum tension on the guidelines. Another procedure is to run the stack and riser with 3 to 6 kips (depending on the water depth, current and weather) until the stack is two or three joints off bottom. Then, increase the tension to maximum for landing the stack. When landing the riser or the stack, maximum tension is recommended.

When the riser is disconnected because of weather and is hanging just a few joints above the stack, maximum tension should be maintained. It is better to have the guideline funnels move up and down on the guidelines than to have the guidelines bend back and forth at the guide posts or loop around some object, as often happens with low tension. Even if the ship is directly over the hole, the riser will probably be displaced by current, and it does not have the weight of the stack to help hold it down. When the lines are properly tensioned, the guide funnels can slip up and down the lines without severe wear. This brings up another point.

Be sure that all guide funnels and guide posts are internally smooth and that they are flared. The most frequent problems are weld burrs and slag that may stick inside of the cones. Be sure to inspect all guide frames after each welding job.

When the guidelines are not being used to run equipment, the tension may be slacked-off to about twice the weight of the line in sea water. Weights of standard size wire rope in seawater are shown in Table 6-1.

Guidelines should be slipped and perhaps cut after each well, depending on the relative water depth of the wells. The same length of wire should not be reeved on the sheaves for more than one well.

Sheave diameters should be about 30 times the outer wire diameter. This is not always practical and is the reason for frequent slipping and cutting of the lines.

With proper care and use as discussed above, guideline breakage will be minimized.

Reentry

One of the problems created by deepwater drilling and subsea completions is relocating and reestablishing communications with the subsea wells. Without guidelines that physically force the equipment to the well, reentry requires that the equipment be positioned for landing by moving the vessel. To be able to position the vessel, the accurate position of the well with respect to the reentry equipment must be determined. Land based navigation equipment is used to position the vessel in the vicinity of the well. Then, the reentry string is lowered to a safe position above the well. Information from TV, often supplemented by acoustic equipment, is used to position the vessel so that reentry can be completed. Accuracy of the navigational equipment and clarity of the water will determine if acoustic devices will be required to supplement the TV camera.

Reentry is used for remotely reestablishing guidelines[12] and on dynamically positioned vessels that do not deploy guidelines.[13,14,15,16] In the latter case, reentry is used for running drilling equipment into the well before the riser is run, for landing the BOPs and for reconnecting the riser when necessary.

Thus, in the slang of offshore oil operations, *reentry* means reestablishing broken communications with a subsea well even when nothing has been entered.

Three tools used for reentry are listed below. In some cases, all three devices are used to supplement each other. They are:

1. Television
2. Acoustic devices for active targets
3. Acoustic devices for passive targets

Television is very important and should be used with the sonic devices. Subsea vision is limited, and acoustic devices are needed for position determination until the string is brought within visual range of the well.

Acoustic devices have the advantage of being able to "see" 500 feet or more through even muddy water; however, the *signature* (picture) seen on a scope can be deceptive. We are accustomed to visual signatures of an object where brightness or darkness depends on reflection and absorption of light. Acoustic signatures depend on sound reflection or absorption and may or may not exhibit the same picture as a visual observation. *Resolution*, being able to distinguish objects that are close together, is a minor problem, but is inherent in sonics.

Acoustic devices employing *active targets* are usually included with dynamic positioning equipment. The targets are acoustic beacons on the well and on the reentry string. Position of the wellhead and the reentry string are determined by the position referencing equipment on the vessel. These positions are displayed relative to each other on a scope.

Some dynamically positioned vessels are equipped with a reacquisition mode that automatically positions the vessel so that the reentry string is nearly over the well. Final connection is usually made while observing the operation on TV.

The major advantages of using active targets are that the targets are well defined and will not be confused with other objects. Also, this method is a natural extension of acoustic position determining systems. The disadvantages are that a beacon requires energy, and consequently has a limited life. Also, the equipment can "see" only active beacons. Thus, obstacles such as sunken ships or irregular seabeds that may be in the path of the reentry string are not detected.

The acoustic device commonly used to detect *inactive targets* are scanning sonar systems that operate like radar. Acoustic energy is emitted from a rotating scanning head. Objects in the energy path reflect the sound back to the scanner, and relative positions of the objects are displayed on a scope. Since the system can "see" not only the well, but other objects as well, a specific signature is important so that the well can be recognized. Signatures can be designed using reflectors in a specific configuration. This technique has been used in 13,000 feet of water by the Glomar Challanger.[16] For final reentry, a TV camera is still recommended.

At one time it was believed that jetting water or mud through an orifice to move the end of the reentry string was advisable. Industry experience has shown that moving the vessel so that the reentry string is directly over the well is usually simpler and preferable to jetting.

References

1. Fisher, W., and Ludwig, M, "Design of Floating Vessel Drilling Riser," *Jour. of Pet. Tech.* (March 1966). p. 272.

2. Tidwell, D.R. and Ilfrey, W.T. "Developments in Marine Drilling Riser Technology," ASME Pet. Mech. Engr, Conf., Paper 69PET14. Tulsa, June 1969.

3. Cook, D., "Floating Riser Helps Extend Drilling to 1700 Foot Depths," *Offshore* (June 1971).

4. Kopecky, J.A. "Drilling Riser Stress Measurements," ASME Pet. Mech. Engr. Conf., Paper 71PET1. Houston, September 1971.

5. Lubinski, A., "Maximum Permissible Dog Legs in Rotary Boreholes," *Jour. of Pet. Tech.* (February 1961). p. 175.

6. Sheffield, J.R., and Caldwell, J.A., "Field Method for Determining Tensioning Requirements in Marine Risers," ASME Pet. Mech. Engr. Conf., Paper 72PET32. New Orleans, September 1972.

7. Crandall, S.H., and Dahl, N.C., *An Introduction to the Mechanics of Solids.* McGraw-Hill, New York, 1959.

8. Childers, M.A., Hazelwood, D., and Ilfrey, W.T., "Marine Riser Monitoring with The Acoustic Ball Joint Angle Indicator." *Jour. of Pet. Tech.* (March 1972). p. 337.

9. Hansford, J.E., and Lubinski, A., "Maximum Permissible Horizontal Motions of a Drilling Vessel," OTC paper 1076. Presented at the Offshore Technology Conference, Houston, 1060.

10. Moore, P.L., *Drilling Practices Manual.* Petroleum Publishing Co., Tulsa, 1974. pp. 220-227.

11. Perry, J.H., *Handbook of Chemical Engineering.* McGraw-Hill Book Co., New York, 1950. p. 390.

12. Sheffield, J.R., and Masonheimer, R.A., "An Acoustic-Mechanical Method for Reestablishing Communications with Subsea Wellheads," *Jour. of Pet. Tech.* (October 1974). p. 1075.

13. Anderson, F.E., "Drilling Reentry without Guidelines," *Proceedings of OECON.* 1968. p. 237.

13. "Deep Ocean Reentry—A Scientific First," *Oil & Gas Jour.* (January 18, 1971). p. 28.

15. "SEDCO 445 Reenters Test Well in 1170 Feet of Water off Borneo," *Offshore* (May, 1973).

16. Sims, D.L., "Acoustic Reentry and the Deep Sea Drilling Program," OTC paper 1390. Presented at the Offshore Technology Conference, Houston, 1971.

7

Motion Compensation

In this discussion, *motion compensation* means compensating for the heave of the vessel. In other words, suspending a load from a floating platform and holding it motionless with respect to the seabed. On the floating platform, this means holding a constant upward force on a reciprocating load.

In the early 1960s, weights called *counterbalances* were used for tensioning both the riser and guidelines. Counterbalances are still used for tensioning guidelines on some vessels, but as drilling moved to deeper water and into areas of severe environmental conditions, counterbalance weights for riser tensioning became so large that they were impractical to handle on a vessel.

Counterbalances were replaced by tensioners that increased the tensioning capacity relative to space requirements. They also facilitated changing tension, yet decreased the deck loading for the same tension. The same principle was applied to keeping a constant weight on the drill bit by holding a constant tension on the drill string.

Another way to hold constant weight on a bit is to use bumper subs (slip joints) in the drill string. As long as the subs are partially collapsed, the bit weight will be nearly constant.

Three approaches will be discussed for motion compensation:

1. *Passive systems* are the most popular. They are used for guideline tensioning, riser tensioning, heave compensators for the drill string, rough weather docking (workboat to semi), and are expected to be used on offshore cranes in the near future. Passive compensators do not require continuous energy input. They use a pneumatic (air) spring to absorb and return energy expended by the motion of the load. Force variations of about 15% of the load can easily be obtained under relatively severe conditions when proper attention is paid to design and plumbing. Much lower force fluctuations are possible, but with increased effort and cost.
2. Active and semiactive systems were experimental in 1979. *Active systems* require external energy to maintain the work required through

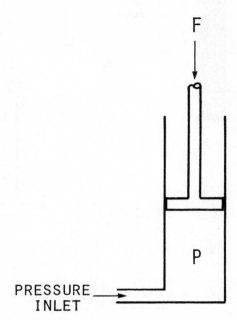

Figure 7-1. A version of the filling
station car rack.

each load cycle. This is done by pumping hydraulic fluid from one
point in the system to another. High initial cost, high instantaneous
power requirements, high prices of energy, and complexity of the
system have limited industry's interest. *Semiactive systems* are hybrids
between the active and passive systems. Continuous energy input is
required, but not in the quantities required by an active system.
3. *Bumper subs* are telescopic joints inserted in the drill string to allow
   for vessel heave. Bumper subs for drilling are not recommended in
   severe weather areas; however, in many areas with occasionally high-
   sea states, bumper subs can be used effectively in floating drilling
   operations.

Passive Motion Compensation

Passive motion compensation includes tensioners and drillstring heave
compensators in common use. The general difference is that tensioners
tolerate a larger force fluctuation of up to 15% about the mean load, while
heave compensators strive for as small a load fluctuation as practicable.

Passive motion compensation uses the principle of the pneumatic spring, a
dynamic version of the well known car rack at a service station (see Figure
7-1). Air pressure exerts a force on a piston that supports a load. Now, if

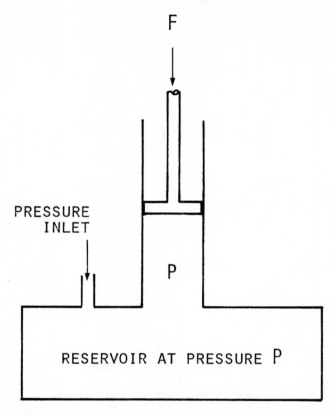

Figure 7-2. A reservoir added to decrease pressure changes.

we allow the load to fluctuate, the force exerted by the gas on the piston
will change. This change in force will depend on the volume of the cylinder
and the volume displaced by the piston. If a large reservoir is added, (see
Figure 7-2) the pressure fluctuations and consequently the force fluctuations
on the piston decrease.

Expressed mathematically:

$$\frac{F_2}{F_1} \propto \frac{P_2}{P_1} = \left(\frac{V}{V + \Delta V}\right)^n$$

where:

P_1 = absolute pressure when piston is at the center of the stroke.
P_2 = absolute pressure when piston has moved so that the volume
ΔV has been displaced.

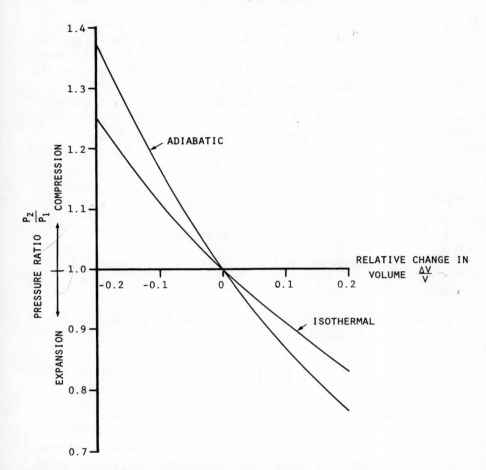

Figure 7-3. Pressure variations with relative changes in volume.

$\dfrac{F_2}{F_1}$ = force on piston related to force when piston is centered.

V = volume of the system with piston centered.

ΔV = volume change caused by piston movement.

n = gas expansion coefficient:
     $n = 1$ for isothermal expansion
     $n = 1.41$ for adiabatic expansion

Since ratios are used, any consistent units will cancel.

Figure 7-3 shows how pressure varies with relative changes in volume (piston movement). The operation is neither adiabatic nor isothermal, and the volume-pressure fluctuations will be somewhere between the two lines.

Theoretical reservoir requirements for a weightless, frictionless, no-flow system are shown in Figure 7-4. The curves were derived from the previous equation, using pressure ratios (P_2/P_1) of 1.00 to 1.30. The compression cycle was used to be conservative. The figure shows how increasing the volume decreases pressure variations.

Let us take a typical case for a pressure fluctuation of $\pm 15\%$ at maximum stroke. From Figure 7-4, this will require a reservoir volume about five times the cylinder volume (adiabatic) Now:

$$\text{reservoir volume} = R \; m \; A \; L$$

where:

R = theoretical $\dfrac{\text{volume of reservoir}}{\text{volume of cylinder}}$ from Figure 7-4

m = number of cylinders
A = piston area, ft^2
L = piston stroke, ft

Consider a 14-inch piston with a 10-foot stroke. We shall hook two units (cylinders) to the reservoir:

$$A = \left(\frac{14^2}{12}\right) \frac{\pi}{4} = 1.07 \text{ ft}^2$$

$$\text{reservoir volume} = 5 \times 2 \times 1.07 \times 10 = 107 \text{ ft}^3$$

Theoretical force variations as a function of piston position and heave can be determined by a modification of the previous pressure-volume equation:

$$E = \left(\frac{F_2}{F_1} - 1\right) 100 = \left[\left(\frac{V}{V - m\, A\, \Delta L}\right)^n - 1\right] 100$$

and

$$-\frac{L}{2} \leqslant L \leqslant \frac{L}{2}$$

where:

E = theoretical percentage error in force applied to the load
$\dfrac{F_2}{F_1}$ = force applied to the load relative to the force with the piston centered

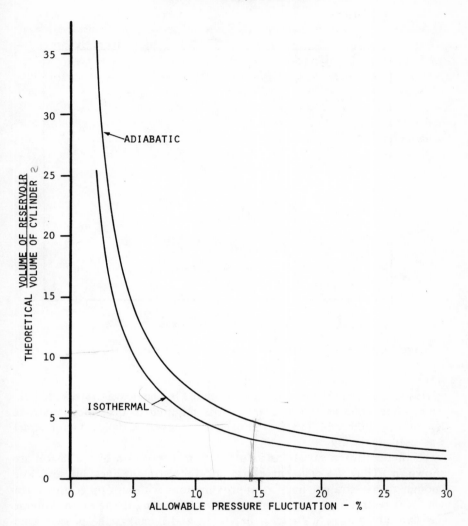

Figure 7-4. Theoretical reservoir size vs. pressure fluctuation.

2 CYLINDERS

← ½ volume of cylinder = ΔV = volume change caused by piston movement

166.5 gal

A = piston area, ft^2

L = full piston stroke, ft

V = reservoir volume + $\dfrac{mAL}{2}$, ft^2

ΔL = length of the stroke from the center of the piston, ft. Never greater

   than $\dfrac{L}{2}$

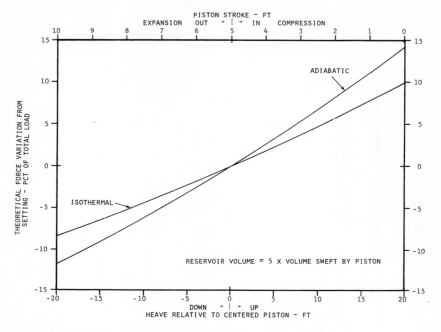

Figure 7-5. Theoretical force vs. piston travel and heave.

Heave relative to the piston travel is related by the heave to stroke length. This is the same as the mechanical loss required to decrease the stroke length, i.e. number of lines reeved over the sheaves. A typical factor of 4:1 was used in the figure.

Theoretical force variations as a function of position of the piston are shown in Figure 7-5. The curves are not quite straight lines, but are close enough to use straight lines as approximations without appreciable error with *this* reservoir to piston volume ratio. Increasing the reservoir volume has two advantages: it flattens the curves and it makes the process tend toward the isothermal line.

In practice, the weight of the piston, rod, hook, lines, sheaves and all moving parts of the motion compensator system increase the pay load, F_l. Sometimes, these weights are neglected.

Friction, inertia and resistance to fluid flow tend to make the curves steeper, increasing the variation of force with heave.[1]

Tensioners

In floating drilling, *tensioners* are used to hold tension on guidelines and on the riser (see Figure 7-6). In the figure, the lower sheaves are attached to

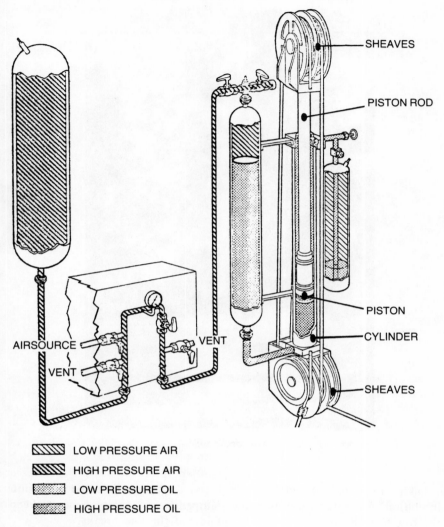

Figure 7-6. Typical tensioner.

the cylinder, the upper sheaves are attached to the rod, and the piston rod applies a force tending to separate the upper and lower set of sheaves.

This separating force determines the tension in the line reeved on the sheaves. Tension is maintained by pressure transmitted to the piston face by oil that is pressurized by the air reservoir. To increase the tension, air is added to the reservoir through a line from a compressor; to decrease the tension, air is vented from the reservoir.

Figure 7-7. Vetco riser tensioner.

Oil is used on both sides of the piston for lubrication and corrosion inhibition. All systems use hydraulic dampening as a safety device to keep the rod from being shot from the cylinder if the line breaks.

Reeving varies from four to eight turns, and decreases the stroke length relative to the heave. For economy, two tensioners may be connected to the same reservoir. Only tensioners that are diametrically opposed on the riser should be connected to the same reservoir. This decreases the probability of cocking the slip joint or tensioning ring while tension is being adjusted or in the event of a reservoir failure. Maximum working pressure of the type of system shown in Figure 7-6 is about 1500 psig. Manufacturers with equipment in the field are NL Rig Equipment (Rucker-Shaffer) and Western Gear.

Vetco also makes a popular tensioner, but arranges the sheaves so that the piston rod is in tension instead of compression (see Figure 7-7). The piston is smaller, and so is the area acted on by the hydraulic fluid. Accordingly, this system uses a higher pressure of 2,500 psig maximum. Lower displacement, however, allows the manufacturer to use a smaller air reservoir that is housed in the structural members, making a more compact unit than the lower pressure systems.

In field operations, four to eight tensioners may be used for riser tensioning. Smaller tensioners varying from 7 to 22 kip capacities are used for tensioning guidelines.

Heave Compensators

A *heave compensator* should apply a constant force to a dynamic load. This allows relative motion of the vessel to be isolated from the load. For example, if we are about to land the stack with a perfect heave compensator, lowering the traveling block 1 foot relative to the rig floor would lower the stack 1 foot relative to the wellhead, regardless of the heave of the vessel.

Of course, there is no *perfect* motion compensator, but the equipment to be discussed allows drilling operations to continue in relatively high sea states by decreasing down time, bit wear and equipment damage while increasing personnel safety.[3] The need for heave compensation was recognized at the advent of floating, but operational equipment was not available even for testing until the turn of the decade, 1960-1970.

More stringent tolerances are placed on heave compensators than on tensioners, because one of their principal uses is to keep a nearly constant weight on the bit even in rough weather. Suppose that 40-kip bit weight is desirable on a bit when the total drill string load is 200 kips. Information by Galle and Wood[2] indicates that drilling efficiency does not deteriorate rapidly until fluctuations exceed about ±30% of the weight on the bit. Using this criterion, fluctuations should be less than 12 kips, which is 6% of the total drill string weight.

Lower fluctuations are preferable, of course. Tolerances less than ±6% are considerably more difficult to hold than the ±15% required of the tensioners (see Figure 7-4). Tolerance of one to six percent is a severe test of the passive motion compensation principle, and is the major reason that semiactive heave compensators have been considered.

Various companies have passive compensators in the field and a discussion of the field operation of different heave compensators was presented by Woodall-Mason and Tilbe.[3] Traveling block and crown block compensators refer to the two different categories of drillstring compensators. They will be discussed separately.

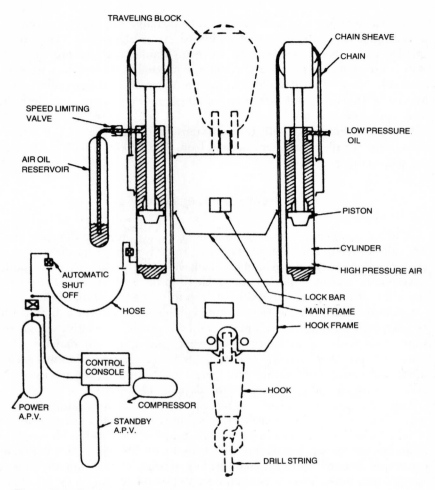

Figure 7-8. Rucker heave compensator.

Traveling Block Compensators

Traveling block compensators locate the operating cylinder(s) on the traveling block, between the block and the hook. Large air reservoirs are located on the deck or below, and pressurized fluid is piped to the cylinder(s) by flexible hoses. Dual or single pistons may be used to support the load. Traveling block compensators are manufactured by NL Rig Equipment (Rucker Shaffer), Vetco, and Western Gear.

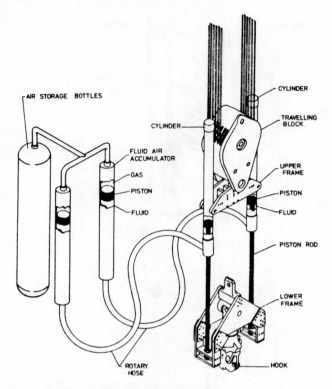

AIR STORAGE BOTTLES

CYLINDER

CYLINDER

TRAVELLING BLOCK

FLUID AIR ACCUMULATOR

UPPER FRAME

GAS

PISTON

PISTON

FLUID

FLUID

PISTON ROD

LOWER FRAME

ROTARY HOSE

HOOK

Figure 7-9. Vetco heave compensator.

The *Rucker* system (see Figure 7-8) uses air on the piston face and low pressure oil dampening on the rod side. Use of air appreciably decreases the friction losses in the hoses between the reservoir and the cylinders, but air in the flexible lines requires special precautions. These precautions include high velocity shut-off valves, and wire rope lines run inside of the flexible hoses and attached to the fittings. These lines keep the hoses from whipping about if they are broken while under pressure. Rucker theoretical data indicate force fluctuations are kept very low by use of gas.

Rucker uses a chain drive to absorb lateral motion of the hook. This lateral motion could cause binding of the pistons if the coupling were rigid.

Vetco makes dual piston (see Figure 7-9) and single piston compensators. *Western Gear* makes a single piston model only (see Figure 7-10). In early tests, the dual piston systems tended to bind from the lateral motion of the hook, but this problem was solved by piston balancing and non-rigid coupling of the rods to the hook frame. Both systems support the load with

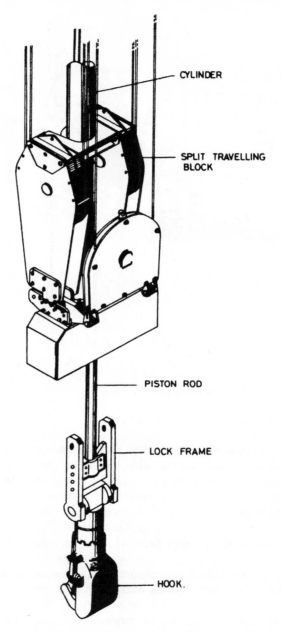

Figure 7-10. Western gear heave compensator.

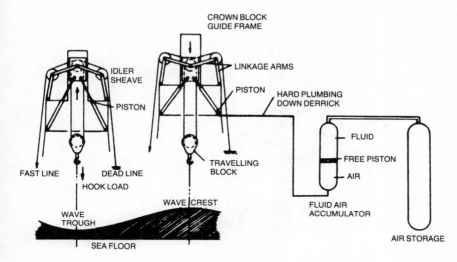

Figure 7-11. Ocean Science and Engineering heave compensator.

high-pressure hydraulic fluid from the remote reservoir to the rod side of the piston. Low-pressure hydraulic fluid on the piston face is throttled for dampening.

Single-cylinder compensators do not encounter problems with lateral loading, but they do require split crown and traveling blocks. The crown block and traveling block must be replaced if the derrick was not initially designed for the single-piston compensator.

Hydraulically operated pistons can be locked in any position by remotely controlled valves that block the relatively incompressible hydraulic fluid in the cylinder. Air operated compensators must be closed and mechanically locked because of the highly compressible fluid in the cylinders.

Crown Block Compensators

Crown block motion compensators are pneumatic springs, but the mechanical compensation is accomplished by taking in and paying out on the *fast* and the *dead* drilling lines. This is done by moving sheaves attached to the piston rods and idler sheaves positioned on arms as shown in Figures 7-11 and 7-12.

The cylinders are located on the crown block and remain essentially motionless relative to the vessel; therefore, rigid piping can be used, decreasing the probability of hydraulic line failure relative to flexible lines required for traveling block compensators. The equipment is massive, and

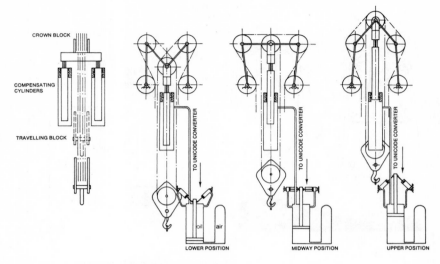

Figure 7-12. IHC heave compensator.

tends to raise the center of gravity of the vessel slightly. More significant is the increased wind loading high on the structure. This increased loading increases the wind heeling moment of the vessel.

Crown block motion compensators have additional geometric effects that are caused by spreading the idler sheaves. By careful design configuration, these geometric effects can be used to flatten the response curve for a motion compensator.

Ocean Science and Engineering (OS&E) uses a center track to guide the tensioning sheaves that balance the force from the cylinder against the line tension (see Figure 7-11). The angle of the cylinder and the length of the arms between the tensioning sheave and the idler sheaves are designed to flatten the curve.

IHC Gusto employs vertical cylinders to maintain tension in the lines (see Figure 7-12). Flattening of the response curve in the area of interest is achieved by a *uincode converter* that actually changes the volume of the liquid reservoir.

Active and Semiactive Compensation

Active compensators are conceptual at present. They are appreciably more complex and require considerably more energy to operate than the passive systems. A schematic diagram of an active system design is shown in Figure 7-13. It is a *closed loop feedback* (see Chapter 3) system that operates

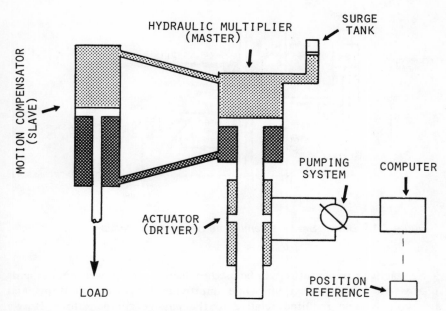

Figure 7-13. Active heave compensator.

in a manner similar to dynamic positioning. As the load reciprocates, a computer-controlled metering pump moves a nearly incompressible fluid from one side of a piston in the driver or actuator cylinder.

The drive cylinder is coupled through a hydraulic multiplier (master cylinder) to the slave cylinder that supports the load. Vessel motion may be sensed by pressure fluctuations in the system through a line attached to the seabed or through other sensing devices. The pump is controlled by the same feedback logic briefly described (see Chapter 3) and the fluid pumping rate will be proportional to the heave rate of the vessel.

On the upward heave, only the energy required to overcome friction losses and to reverse the inertial components of the system is required. During the downward heave of the vessel, the compensator must lift the entire load. For example, the peak horsepower required for a 400,000-pound load on a vessel heaving downward at four feet per second would be (without friction or mechanical loss):

$$\text{maximum horsepower*} = (400,000) \ (4)/550$$
$$= 2.9 \times 10^3 \text{ Hp surge}$$

*Note numerical value of mass/force conversion is one (1). This is an appreciable fluctuation of electric load even on a drilling vessel.

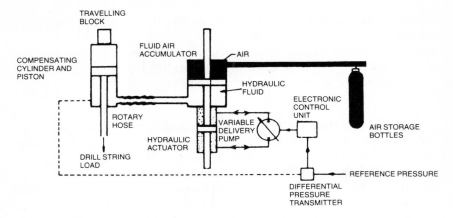

Figure 7-14. Brown Brothers semiactive heave compensator.

Semiactive compensators can be used to decrease the power requirements relative to an active system while improving the force variations and reservoir volume required relative to the passive compensators. Brown Brothers has a semiactive compensator design that combines the active and passive systems (see Figure 7-14). This might be called a passive system with power injection from a servo controlled pump. The system is designed for a 400,000-pound load with variations of ±2500 pounds (0.63%) when the heave is 5 feet. Power requirements are 250 horsepower and air storage requirements are estimated to be about 60% of the requirements of a "comparable passive system."

Bumper Subs

Bumper subs are telescoping joints inserted into the drillstring to decrease the transfer of vessel heave to the bit. Bumper subs are required to transfer torque to the bit—a tough requirement. There are two general types of subs: balanced and unbalanced. *Unbalanced bumper subs* are typical telescoping joints where internal pressure operates on the end of the wash pipe (inner pipe) and the annular pressure operates on the outside area of the case. The result is an opening force on the sub.

Balanced bumper subs have special internal chambers and porting that equalizes the areas acted on by the internal and annular pressures.

Extensive design work was done on bumper subs prior to the advent of heave compensators, because bumper subs were the means for keeping a nearly constant weight on the bit when drilling from a floater. Only a brief summary will be included here.

Bumper subs fit into the drill collar string, but do not have the ruggedness of drill collars. To obtain a reasonable operating life from bumper subs, action has to be taken on the following points of concern:

1. Always run the bumper subs at the neutral point in the string.
2. Bumper subs are designed to stroke, and if operated at a single position, they will wear at that position.
3. Running the subs in tension minimizes the area through which the torque will be transmitted, and will cause excessive wear and fatigue.
4. Running the joint in compression increases torque reversals and fatigue in the tool. It also increases "bit chatter" and instantaneous peak torque, causing unnecessary wear to the tool and the string. However, this is the fault of the operation, not the sub.
5. From items 3 and 4 above, it is evident that once the string has been run, the weight on the bit is fixed. This weight should not be changed until the string has been pulled and the number of drill collars below the bumper sub changed.
6. Do not separate the bumper subs in a drillstring for heavy drilling operations.
7. Special care and maintenance are required to obtain a reasonable operating life.

Remember that bumper subs fit into the same part of the drillstring where drill collars and stabilizers have been destroyed. This is a severe test of workmanship and maintenance.

Bumper subs should be maintained and operated in accordance with the manufacturer's specifications. Replacing seals, worn parts and lubricating grease is important. Make-up torque when reassembling the subs is very important.

Special maintenance shops (see Figure 7-15) designed to handle such equipment are available. The shops are compact, and contain hydraulically operated tongs that are useful for maintaining jars and other fishing tools as well as bumper subs.

Bumper subs should be locked closed when they are being handled on the rig floor. Picking up bumper subs and allowing them to collapse when they are "stood back" in the derrick is a hazardous practice that should not be permitted on a floating drilling rig.

Figure 7-15. Special maintenance shops for bumper subs and fishing tools.

References

1. Kozik, T.J., and Noerager, J., "Riser Tension Force Fluctuations," OTC paper 2648. Presented at the Offshore Technology Conference, Houston, 1976.
2. Galle, E.M., and Wood, H.B., "Best Constant Weight and Rotary Speed for Rock Bits." Paper presented at the spring meeting Pacific Coast District, Production Div., API, Los Angeles, May 1963.
3. Woodall-Mason, N., and Tilbe, J.R., "Value of Heave Compensators to Floating Drilling," *Jour. of Pet. Tech.* (August 1976) p. 938.

Formation Testing

The purpose of *formation testing* is to obtain as much information as possible about the flow characteristics of the reservoir and the reservoir fluids so that the formation can be evaluated economically. This includes accurate transient pressure data, accurate determination of flow rates of all formation fluids and physical properties including the PVT* relationships of these fluids.

In an exploration well, the final testing and formation evaluation is a major reason for having drilled the well in the first place. Completion events—especially cementing of the production casing or liner—which lead to testing are important. Naturally, planning of the formation test should not be slighted.

All available information should be used to estimate the formation characteristics and fluid properties. A close estimation of the fluids to be produced and pressureflow relationships to be encountered is important in selecting the preferred equipment from that which is available to do the job and control the flow rates within the required limits. The better the planning, the more likely the earlier tests are to be successful. Familiarity and proper use of the equipment are very important in obtaining valid test data early.

Well testing from a floater requires special equipment to compensate for vessel motion, and special subsea equipment to shut in the well if the vessel has to disconnect from the wellhead. The necessary additional equipment and redundancy complicate the test string and the operation, leading to increased problems with equipment and data interpretation. Such complications combine with delays caused by weather to make formation testing from a floater expensive, and increase the incentive for improvements in equipment and planning.

Based on the above information, these major safety precautions should be followed:

1. Always set the test packer in the casing or in the liner, never in the open hole.

* *PVT* means pressure-volume-temperature

2. Perform testing in a liner or casing.
3. Use a test tree in the BOP stack. Include an emergency downhole shut off near the formation. Safety devices will be discussed with the equipment.
4. Stop testing if any of the following danger signals occur:
   a. Pressure approaches the pressure rating of any of the equipment used.
   b. Equipment becomes overloaded by excessive production.
   c. Annular pressure indicates communication between the annulus and the test string.
   d. Gases reach explosive levels in critical areas, or poison gases are produced.
   e. Vessel motion approaches unsafe conditions.

As on land, it is advisable to commence flow during daylight. Releasing the packer is preferable during daylight also.

The downhole equipment will consist of a packer, a test tool (valve) to open or shut in production near the formation, and reverse circulating subs to reverse out the produced fluids. Included also will be various sensors, chokes, samplers, jars and special downhole shut off valves. This downhole assembly will be connected by tubing or drillpipe to a mudline shut off.

Downhole Test Equipment

It would be nice to be able to detail the downhole test assembly for all occasions. Testing, however, is like drilling a well: it requires separate procedures and some variation in equipment to meet the existing situations as they arise. The local service contractors, their technical ability, and the tools that will be available are influential in determining the test string design.

Drillpipe that is in good condition may be used for the test string, as may tubing. Drill collars may be needed for adding weight to different parts of the string, depending on the equipment used.

Test Tools

The test tool may have to be opened and closed more times on a floating rig than on a land rig because of the precautions required. Therefore, the two flow period test tools that may be used on land are generally inadequate for testing from a floater. Two basic types of multiple flow test tools are currently used for testing from a floating drilling vessel. Test valves operated by annulus pressure are relatively new, and may not be available in all areas of the world. These tools operate by pressure only, and do not require drillstring manipulation.

The second type of test valve is operated by reciprocating the test string. By 1977, rotating test tools had generally disappeared from floaters.

Tools actuated by increasing annular pressure. These valves are activated by applying 1,000-1,500 psi pressure to the annulus. Releasing this pressure allows the valve to close. Additionally, pressuring the annulus to 2,000-2,500 psi will close and lock the valve. This traps a sample of reservoir fluid after the final flow period. The valve also is a safety device for shutting in the well if the tubing parts and casing pressure increases above 2,500 psi. Available valves are the Halliburton Annular Pressure Operated (APO) valve, the Halliburton Annular Pressure Regulated (APR) valve and the Johnston Pressure Controlled Tester (PCT) valve. The APO and PCT valves are sliding sleeve valves. The APR valve is a ball valve that has a larger flow area than sliding sleeve valves. The actuation mechanisms in these valves are similar, so the operation of the APO only will be discussed.

Valve operation is shown schematically in Figure 8-1. A piston mounted on a sliding sleeve valve balances the annulus pressure against the pressure in an inert gas chamber. At annulus pressures below about 1,000 psi, the ports in the valve are closed to flow. Application of about 1,000 psi pressure moves the valve mandrel upward as shown in the figure, allowing flow of fluids through the tubing. Pressure build up on the order of about 2,000 psi will force the valve mandrel past the flow position into a locked closed position. Once the valve is in the locked closed position, it cannot be reopened until it is brought to the surface.

Reciprocating test tools. The *Johnston Multiple Flow Evaluator (MFE)* is operated by reciprocation and requires a full cycle for actuation. When the tool is first set down on the packer, the test valve will open after about a three-minute to five-minute delay. Picking up and setting down will shut off the flow. Each alternative reciprocation cycle will open or close this valve. The number of times the MFE can be opened and closed is unlimited.

The *Halliburton Hydrospring Tester* is opened by compression and closed when the compression is removed. It is advisable to use about 20,000 pounds of weight for satisfactory valve operation. Drillpipe or tubing, or a shorter length of drill collars, may be used to make up this weight.

Slip Joint Safety Valve

The *slip joint safety valve* opens when it is in tension and closes when the slip joints are completely collapsed. This shuts the test string below the mudline if the string should part. The valve has a sliding sleeve mechanism (see Figure 8-2) attached to and running through the slip joints. The valve mandrel is attached at the top of the upper slip joint and the valve sleeve is

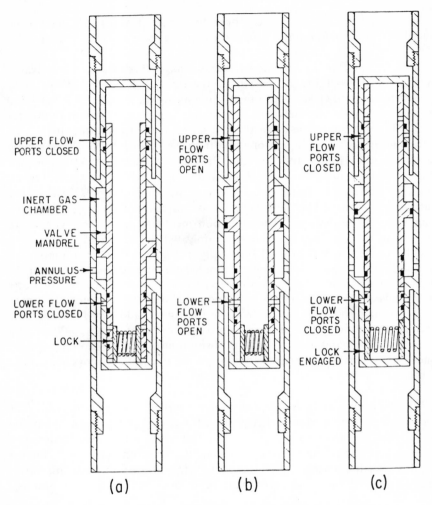

UPPER FLOW
PORTS CLOSED

INERT GAS
CHAMBER

VALVE
MANDREL

ANNULUS
PRESSURE

LOWER FLOW
PORTS CLOSED

LOCK

UPPER
FLOW
PORTS
OPEN

LOWER
FLOW
PORTS
OPEN

UPPER
FLOW
PORTS
CLOSED

LOWER
FLOW
PORTS
CLOSED

LOCK
ENGAGED

(a)   (b)   (c)

Figure 8-1. Example of a test valve operated by annulus pressure.

attached at the bottom of the lowest slip joint in the series. The safety valve is opened when any of the joints are partially extended, but it is closed when all of the joints are closed. The valve may be deliberately closed by lowering the string so that all of the subs are collapsed.

Volume-Pressure Balanced Slip Joints

The *volume-pressure balanced slip joints* are designed to eliminate pressure and flow surges caused by vessel motion. They are primarily used

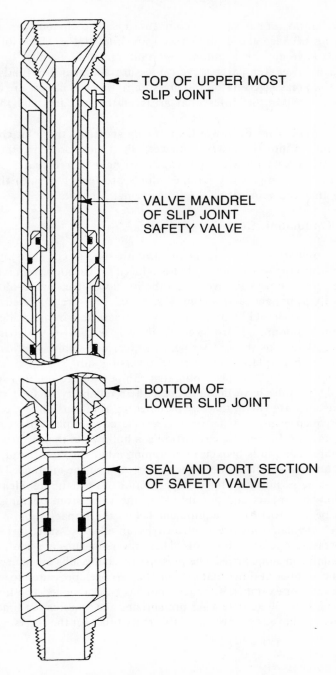

Figure 8-2. Slip joint safety valve.

by vessels that do not have motion compensators: The slip joints are inserted in the string between the drilling vessel and the BOPs. These joints (see Figure 8-3) have an upper annular reservoir/pressure chamber for tubing fluids and a lower annular reservoir/pressure chamber for fluids in the annulus. The areas acted on by the pressure in these chambers are equal to and compensate for the differential areas of the mandrel and the outer sleeve.

As the vessel heaves downward, the string shortens, the joint collapses, and fluid moves into the reservoir to keep the fluid velocity in the tubing constant. As the vessel heaves upward, the string length increases, the joint extends, and the temporarily retained fluids are reinjected into the main stream so that the velocity of the fluids remains constant.

Reverse Circulating Subs

Reverse circulating subs or *reversing subs* are used to circulate mud from the annulus into the test string so that produced fluids can be removed from the string. Four types of reversing subs are available: pump-out plugs, rotating valves, impact, and sliding sleeve and pressure operated.

Reversing subs should be placed as low as practicable in the string so that the minimum amount of produced fluids will remain in the string after circulation. Design of the test string, however, will strongly influence the maximum depth of placement of these reversing subs.

The *pump-out plug* sub has a plug that can be pumped out at a specified pressure. These subs have demonstrated their reliability in field service.

Rotating subs are valves that are actuated by rotation. They are reliable and are frequently used as a primary reverse circulating sub.

The *impact sub* for reverse circulation is a hollow pin that protrudes into the pipe bore. The pin is broken by dropping a bar through the pipe bore, allowing communication between the test string and the annulus. This type of sub must be positioned above all restrictions in the tubing. Heavy mud will cushion the impact and possibly keep the valve from opening. A bar catcher is needed to protect equipment below the impact sub.

Pressure operated subs are controled by annular pressure, just like the annular-pressure-operated test tools. These reversing subs are new and may not be available in some areas. The pressure reversing subs use a mechanism to count the number of times that the annulus has been pressured to a certain pressure level. For example, if the test tool has been actuated five times and the reversing sub will actuate after pressurizing the annulus ten times, then the annulus will have to be pressured five more times for the reversing sub to operate.

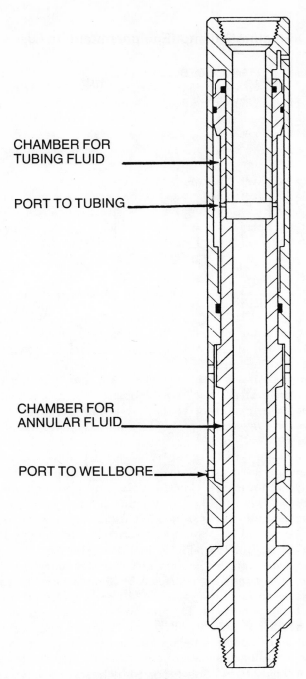

CHAMBER FOR
TUBING FLUID

PORT TO TUBING

CHAMBER FOR
ANNULAR FLUID

PORT TO WELLBORE

Figure 8-3. Volume-pressure balanced slip joint.

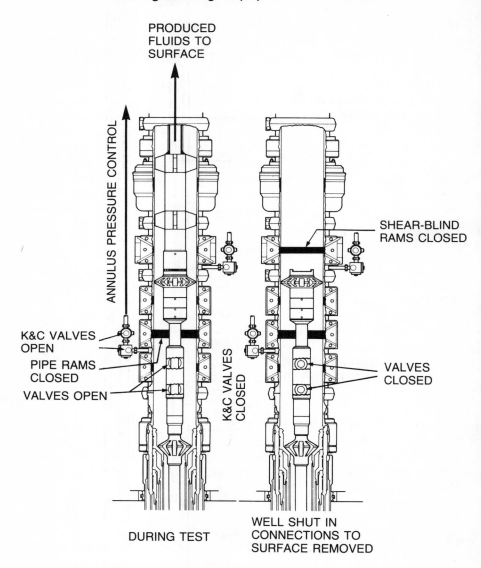

Figure 8-4. Example of seafloor shut-in equipment.

Sea Floor Shut In

Special *trees* are designed to land in the wellhead and blowout preventers, to provide a means of tubing shut-in at the sea floor (see Figure 8-4). These trees use redundant, hydraulically operated valves. One or both will be ball valves, depending on the manufacturer. Hydraulic pressure is furnished through a flexible line from the surface. When the *hold open* hydraulic pressure is released, a spring closes the valves. The valves are designed to shear a wire line if necessary. The pipe to the surface is released hydraulically, and a mechanical back-off may be used in case of hydraulic failure. This allows the vessel to disconnect the riser from the wellhead and have the well shut in. The shear/blind rams may be closed in most current systems; however, some of the earlier systems would not fit below the shear/blind rams, requiring the well to be shut in on the annular preventer.

The newer trees are designed for deepwater, while earlier models relied too heavily on the internal tubing pressure to assist in closing. These earlier models may not close when hydraulic pressure is released in deep water. Valves that might have this limitation should always be evaluated before using.

Hang-off shoulders are provided to support the tree in the rams. A fluted hang-off point that will not pass the wellhead is included. Since spacing between the rams will vary from rig to rig, a tree should be modifiable to fit different stacks, or it should be designed to fit into a specific stack and used with that stack only.

During a test, the rams should be closed around the test tree, and the casing pressure should be monitored through a kill/choke line.

Manufacturers of this equipment are Baker, Otis and Johnston (Flopetrol).

Space Out

Space out for the two types of reciprocating test tools will be discussed here. The first will be for the single point hang-off, such as will be required for the full cycle reciprocating tool or an annulus operated tester. The second will be for a two-point hang-off, to open and close a tester such as the Halliburton Hydrospring tester.

An example test string arrangement is shown in Figure 8-5. The single hang-off is relatively simple because there is only one point to consider for flow or shut off. For example, use the dimensions shown in Table 8-1.

A suggested procedure is to space out so that when the string is hung off, the lower slip joints are fully collapsed, and the upper slip joints are at mid stroke.

The space out procedure can be the following:

1. Land shoulder in wellhead and mark pipe (only the large shoulder will land in the wellhead)
2. Pick up at about 18 feet (shoulder about 4 feet above rams)
3. Close rams and land shoulder by lowering 4 feet
4. Mark pipe

(test continued on page 195)

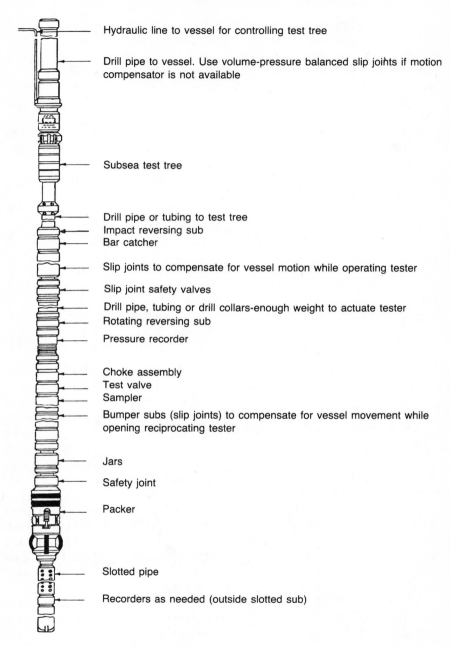

Hydraulic line to vessel for controlling test tree

Drill pipe to vessel. Use volume-pressure balanced slip joihts if motion compensator is not available

Subsea test tree

Drill pipe or tubing to test tree
Impact reversing sub
Bar catcher

Slip joints to compensate for vessel motion while operating tester

Slip joint safety valves

Drill pipe, tubing or drill collars-enough weight to actuate tester
Rotating reversing sub

Pressure recorder

Choke assembly
Test valve
Sampler

Bumper subs (slip joints) to compensate for vessel movement while opening reciprocating tester

Jars

Safety joint

Packer

Slotted pipe

Recorders as needed (outside slotted sub)

Figure 8-5. Example test string arrangement.

Table 8-1
Space out Lengths for Example Test String

Distance from wellhead landing point to landing point on the rams for large shoulder	14 feet
Lower slip joint stroke	12 feet
Tester operation	2 feet
Upper slip joint stroke (safety valve closes in last foot of collapse)	10 feet
Packer: distance to set	1 foot

5. Open rams and pick up 20 feet
   a. Lower slip joint stroke 12 feet
   b. Tester open 2 feet
   c. Packer set distance 1 foot
   d. Half of upper slip joint 5 feet
6. Set packer
7. Lower handoff shoulder and hang off on rams. When the packer is set, the hang off point on the tool will be about 19 feet above the rams.

A weight loss approximately equal to the weight of the string below the upper slip joints will indicate that the packer has been set.

If the Johnston MFE is run, it will open in three to five minutes. If an annulus pressure operated tester is run, annulus pressure of about 1,000 psi would be required to operate it. During a drill stem test flow period, always pressure the annulus to about 1,000 psi and continuously monitor this pressure. A pressure increase indicates communication between the tubing and the annulus, while a pressure decrease indicates a leak. In either case, the test should be stopped and the problem corrected.

For a tester that requires compression to close and removal of compression to open, additional considerations are necessary. Space out depends not only on the closure of the slip joints but also on the combined configuration of the BOP and the test tree. One example would be to space out so that when the upper (larger) test tree shoulder is hung off on a specified ram, the well flows (see Figure 8-6).

When the lower shoulder is hung off on the same ram, the well is shut in (see Figure 8-7). For the well to flow, the lower slip joints must be completely collapsed and the upper slip joints must be partially collapsed. When the well is shut in, the upper slip joints will be partially extended.

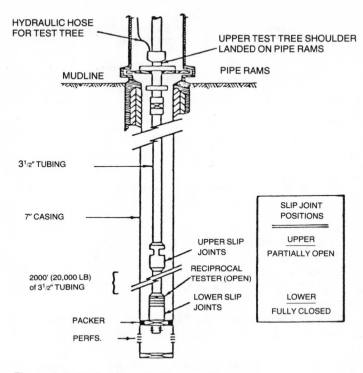

HYDRAULIC HOSE FOR TEST TREE

UPPER TEST TREE SHOULDER LANDED ON PIPE RAMS

MUDLINE

PIPE RAMS

3½" TUBING

7" CASING

UPPER SLIP JOINTS

RECIPROCAL TESTER (OPEN)

2000' (20,000 LB) of 3½" TUBING

LOWER SLIP JOINTS

PACKER

PERFS.

SLIP JOINT POSITIONS

UPPER PARTIALLY OPEN

LOWER FULLY CLOSED

Figure 8-6. Test string—flowing position.

The test tree in the example (Table 8-1) has the distance of eight feet between the upper hang-off shoulder and the lower hang-off shoulder. In this case, the procedure might be:

1. Land the tree in wellhead and mark pipe (only large shoulder will land in the wellhead).
2. Pick up 18 feet (rams midway between test tree shoulder).
3. Close rams and land upper shoulder by lowering string about 4 feet.
4. Mark pipe.
5. Open rams and pick up about 17 feet.
6. Set packer and lower string to about 10 feet below the lower mark.
7. Close rams and land lower shoulder on the rams by lowering the string about 2 feet.

Prior to closing the rams, a weight loss somewhat less than in the previous example would indicate that the packer has been set. The string can be lowered to the opening position to check proper setting and raised without starting flow of the tester, if the tester is not held in compression more than 2½ minutes.

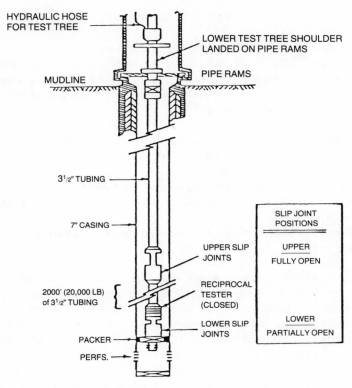

HYDRAULIC HOSE
FOR TEST TREE

LOWER TEST TREE SHOULDER
LANDED ON PIPE RAMS

MUDLINE

PIPE RAMS

3½" TUBING

7" CASING

UPPER SLIP
JOINTS

RECIPROCAL
TESTER
(CLOSED)

2000' (20,000 LB)
of 3½" TUBING

LOWER SLIP
JOINTS

PACKER

PERFS.

SLIP JOINT
POSITIONS

UPPER
FULLY OPEN

LOWER
PARTIALLY OPEN

Figure 8-7. Test string—shut-in position.

Space out is designed so that during the flow period:

1. The lower slip joints are fully collapsed 12 feet
2. The tool is operated 2 feet
3. The upper slip joints are collapsed by 2 feet

This allows the packer to be set within ± 2 feet of the intended location and still have the equipment operate as planned. Larger spacing between the shoulders will allow more tolerance in setting the packer. However, it is important to keep the tree within the BOP so that the tree can be released and the blind rams closed in case of an emergency.

Test trees must be designed for the BOP configuration or they must have assembly features to allow hang-off. Slots in the tree must coincide with the ram configuration at different stages of the test where reciprocating tools are used. A "universally" used tree should be capable of profile modification by changing out nipples on the tree.

Surface Test Equipment

At the surface, the test equipment is similar to equipment used for a land production test. The major differences are that the produced fluids are burned, and the separators should have additional baffles to decrease the sloshing of liquids caused by vessel motion. The precautions used on land are also important in floating drilling:

1. Formation fluids should be flowed through steel lines.
2. Parallel choke runs should be used to allow changing out a choke without shutting in the well.
3. Positive chokes should be used during the flow period.
4. Pressure and flow limitations of the surface equipment must be known.
5. Kill mud should be ready for injection into the test string or annulus when needed.

Use of downhole chokes, water pads and flow rates varies between operators. Usually, the flow is controlled at the surface, and a positive choke (bean) should be used for precise control of the final flow period.

A schematic of the surface equipment hook up is shown in Figure 8-8. The produced fluids are run through a choke, then a heater (if needed), and then through separators. Gases are flared, and produced liquids are retained in a holding tank. From the tank, the liquids are pumped to a burner and incinerated. In some cases, the volume of the holding tank may be relatively small and may act only as a surge tank.

Surface test equipment can be purchased in a package especially designed for a rig, or the individual components can be purchased and assembled separately. Care should be exercised when designing a test system. Changing components in the middle of fabrication can drastically affect the overall efficiency of the system. Not only the major components, but the auxilliary equipment such as pumps, air compressors and flow lines will affect the overall performance of the system.

Heaters are needed for gas and gas-condensate production. Live steam, "diesel fired" or hot water are frequently used for heating the produced fluids. Of the three, live steam seems to be becoming the most popular because it is more efficient than hot water and safer than diesel fired units.

Separator sizing is an art, and it is usually advisable to seek help from persons familiar with this work. Unfortunately, separator efficiency often requires field refinement to operate efficiently. Since production surface equipment experts are usually not available on a floater, special care must be taken in the initial design if the equipment is to work properly.

An example of a horizontal separator is shown in Figure 8-9. In this separator, produced fluids enter at the right and pass through a gas/liquid separator that breaks up the flow steam into a spray to aid gas evolution.

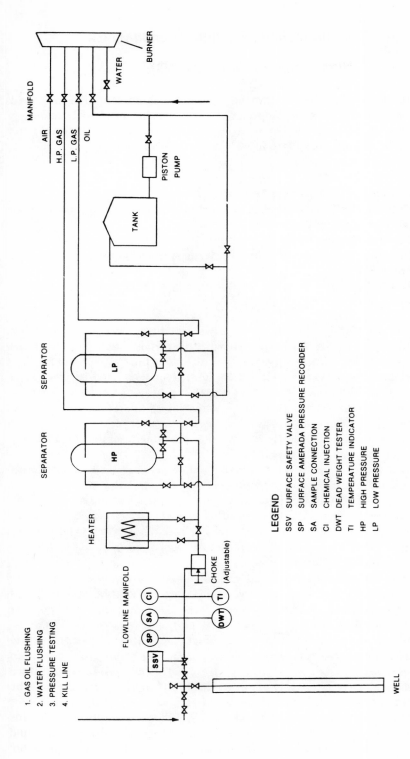

Figure 8-8. Schematic of an oil well test system.

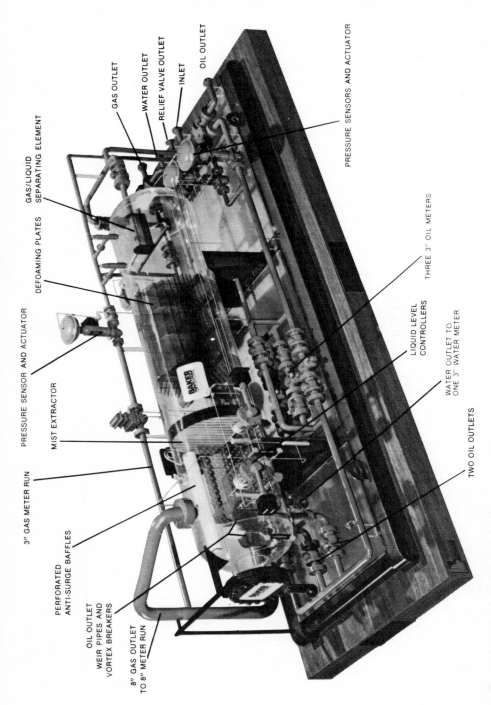

GAS OUTLET

WATER OUTLET

RELIEF VALVE OUTLET

INLET

OIL OUTLET

PRESSURE SENSORS AND ACTUATOR

GAS/LIQUID SEPARATING ELEMENT

DEFOAMING PLATES

THREE 3" OIL METERS

PRESSURE SENSOR AND ACTUATOR

LIQUID LEVEL CONTROLLERS

MIST EXTRACTOR

WATER OUTLET TO ONE 3" WATER METER

3" GAS METER RUN

TWO OIL OUTLETS

PERFORATED ANTI-SURGE BAFFLES

OIL OUTLET WEIR PIPES AND VORTEX BREAKERS

8" GAS OUTLET TO 8" METER RUN

Figure 8-9. Horizontal test separator.

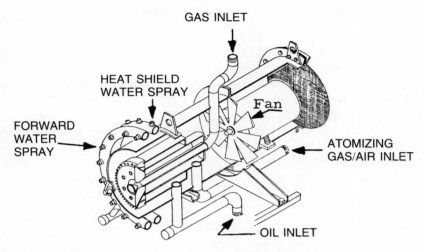

Figure 8-10. Isometric section of a 10,000 BOPD burner.

The gas rises, and the liquids fall to the bottom of the separator. Both gas and liquids pass through defoaming plates. These plates knock some of the liquids drops out of the gas, and also chop up the foam so that the liquids settle, and the gas will escape more readily.

The gas passes through a mist extractor that may be a series of angle irons or wire screens. The objective is to make the gas turbulent so that entrained droplets are knocked out of the gas, coalesce on the metal and run down into the liquids below. Gas exits from the top of the separator and into the gas meter run.

Liquids pass along the bottom and are contained by baffle plates that restrict liquid sloshing. The liquids pass through weir pipes that should be equipped with vortex breakers to avoid entraining gas in the liquid that is withdrawn from the separator and run to the meters. Large lines and small pressure drops are necessary in oil meter runs to avoid additional gas being expelled from the oil. Erroneous readings will result from gas effervessing during oil volume measurements.

Burners may be rated for flow rates as high as 12,000 BOPD with smokeless incineration. These burners consist of a *atomizing head* that emits oil in a small spray by impinging air or an air/gas mixture on the oil stream. This spray passes a shroud through which more air is blown from a close-mounted fan. The oil is ignited by gas pilots, complete with an electric ignition system. Two water ring manifolds are furnished: One sprays forward into the flame to reduce smoke generation; the other sprays nearly vertically from the ignition head as a heat shield in an attempt to shield the rig from the high temperatures of petroleum incineration. A schematic of a typical burner is shown in Figure 8-10.

References

1. Harris, L.M., *An Introduction to Deepwater Floating Drilling Operations.* Petroleum Publishing Co., Tulsa, 1972. pp. 162-177.

2. Christman, S.A., and Masonheimer, R.A., "Drillstem-Test Assemblies for Floating Vessels," *Journal of Petr. Tech.* (August 1974). pp. 851-855.

3. Wray, G.Q., Petty, G.E., and Jeffords, C.M., "Developments in Testing from Floating Drilling Vessels," SPE paper 3094. Presented at the 45th Fall Annual Meeting, SPE of the AIME, Houston, 1970.

4. Nutter, B.P., *et al*, "Safe, Pollution Free Testing of Offshore and Arctic Wells," OTC paper 2871. Presented at the Offshore Technology Conference, Houston, 1977.

Special Problems: Navigation, Workboats, Tides

9

Three items that are important to floating drilling operations do not require an entire chapter for discussion: Location determination, workboats, and Tides. These important items are discussed here under the title of *Special Problems*.

Location Determination

In offshore drilling operations, no survey team can run a traverse to the location, yet the need for accurately determining positions still exists. In oil exploration drilling, accurate reproducibility or repeatability of position data is necessary for relocating potential discoveries determined by seismic exploration. When approaching national or lease boundaries, however, absolute positions on the globe become equally important. Modern techniques of navigation and location determination include use of radio positioning and satellite systems.

Radio Positioning Systems

Radio positioning systems play an important role in marine activities, particularly in regard to survey control for the vast marine geophysical data collection missions.[1] Radio positioning systems use radio (electromagnetic) signals transmitted from stations on shore to permit the mobile offshore station to *fix* its location. Examples of some of the common systems are: HIRAN, LORAC, LORAN, OMEGA, RAYDIST, DECCA, SHORAN, AND X-Band Radar.

Each of these systems uses electronic instruments to provide a geometric grid for position fixing. These grid systems can be divided into four categories:

1. *Circular systems*: Measure distances directly.
2. *Hyperbolic systems*: Measure distance differences.

3. *Circular/Hyperbolic systems*: Combination of the circular and hyperbolic systems.
4. *Hyperbolic/Elliptic systems*: Use both the sum and difference of distances.

Distant measurement. Radio navigation systems measure distances electronically. Distance measurement methods may be referred to as pulsed, as phase comparison, or as a combination of the two.

An example of the pulsed system would be a transponder that responds to a signal from a mobile unit. The mobile unit sends a signal, and the fixed station responds. Knowing the speed at which the signals travel and the turn-a-round time of the transponder, the distance between the stations can be calculated.

A phase comparison system transmits a signal at two different wave lengths simultaneously, and the receiver measures the phase angle between the two sinusoidal waves. Since any given phase angle will recur at equal segmented distances from the transmitter, the operator (or the equipment) must know the approximate distance to the transmitter. The recurring distances are equal length segments along the radius of the radiated energy path and are referred to as *lanes*.

Phase comparison systems may be used in transponder systems. For example, the signals may be pulsed and the time of arrival used to determine the lane, while phase comparison is used to determine the exact position within the lane.

The above methods are used to determine distances. The distances to the different shore stations may be added or subtracted to form different grid patterns for navigation.

Circular systems. These measure the distance from the mobile unit to two or more fixed stations. The fixed stations are located on shore or on fixed platforms. These fixed stations are surveyed and their position accurately known.

With the two fixed stations plotted on a chart, the scaled distance to the stations can be used with a compass to determine position of the mobile unit (see Figure 9-1). The shore stations are commanded from the mobile unit. This limits the number of units that can use the same fixed stations.

Radar, LORAC, SHORAN and RAYDIST are examples of navigation systems that use circular configurations. Current RAYDIST systems provide for multiparty operations and require no transmission from the mobile unit.[3]

Circular systems are usually very accurate, but are limited to short ranges. The accuracy of a circular system depends on having a triangle with reasonably large legs relative to the base line between the two shore stations.

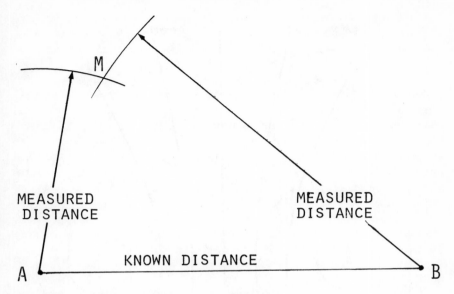

Figure 9-1. Navigation using a circular grid.

For example, assume the stations are on islands and the mobile unit is going to sail between them. As the range circles approach tangency, the position determination becomes inaccurate.

Circular systems were the backbone of early offshore exploration, but are not always practical in extensive offshore operations. Usually, contractors are engaged for specific surveys in given areas and set up the shore stations and man the mobile unit.

Hyperbolic systems. With a circular grid, each shore station is contacted individually, and the distance to the station is determined from this information. Another navigation technique uses a hyperbolic grid.

In a *hyperbolic system*, both fixed stations are triggered simultaneously and the difference in distances to the two transmitters is taken (see Figure 9-2). If the distances are the same, the mobile unit is somewhere on line 1. If the mobile unit is closer to Station A than to Station B, the mobile unit will be located on or between the lines that are shown.

If a third station is added, a grid comparable to that of Figure 9-3 can be made for determining vessel location. The advantage of this system over the circular system is that it can be used by any number of mobile receivers. In Figure 9-3, stations A and B are called *slave stations*, and the unnamed station, referred to as the *master*, transmits at periodic intervals. The slaves merely transpond to this signal, and differences in the signal arrival times are all that are needed to fix the mobile units position within a given lane.

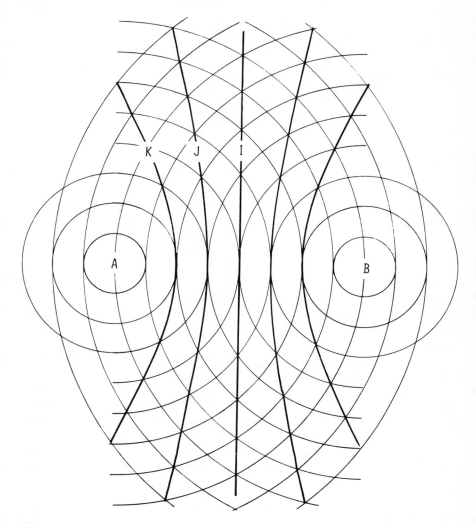

Figure 9-2. An example of a hyperbolic grid formed by taking the difference in distance to the fixed stations A and B.

LORAN (LOng RAnge Navigation) is an example of a hyperbolic navigation system. There are two types of LORAN systems, LORAN-A and LORAN-C. LORAN-A is a pulsed hyperbolic system operating in the 1,800 to 2,000 kHz frequency band. Time difference measurements from two or more station pairs are used to fix a position. Accuracy is from one to five nautical miles, depending on the location of the receiver within the

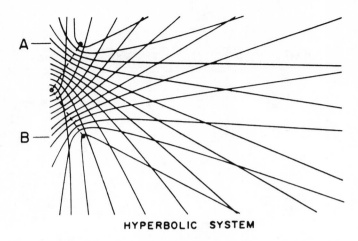

A

B

HYPERBOLIC SYSTEM

Figure 9-3. Grid configuration of the hyperbolic system.

hyperbolic grid. *Groundwave* range over seawater is approximately 750 miles. The LORAN-A is used extensively by the fishing industry. The theoretical one-quarter nautical mile fix accuracy is only realized in limited areas.[4]

LORAN-C has been selected as the government-provided radio navagation system for the Coastal Conference Zone, as announced on May 16, 1974. LORAN-C is a pulsed hyperbolic system operating in the frequency band of 90 to 110 kHz, using phase comparison of the received signals to establish a line of position. Time difference measurements between pulses received from the master and slave stations are used to resolve the lane ambiguities inherent in phase comparison systems (circular/hyperbolic). LORAN-C in the range-range mode offers a considerable advantage, extending the coverage area within which accurate fixes are obtainable.

Previous limitations of LORAN-C were that special equipment was required and that the receivers were not reliable. However, these limitations no longer exist, because of major breakthroughs in technology. The system can now meet and fulfill the quarter-mile accuracy requirement established for the U.S. Coastal Zone.[4]

OMEGA is a world wide navigation system that, when fully operational, will use only eight stations to provide global coverage. OMEGA is a very low frequency phase comparison system that produces a family of hyperbolas whose foci are any two of the transmitting stations.[2,5] Because of the high power of the stations and the low frequencies used, the OMEGA signals propagate over extremely long distances. It is possible to obtain a one mile to two mile fix at ten second intervals anywhere on earth with this system.

Table 9-1

Table of Comparative Radio Positioning System Characteristics[2]

System Type	Frequency	Nominal Range	Nominal Accuracy	Tracking Mode	Relative Crew	Relative Ship. Wt.	Special Problems	Special Advantages
Omega	10.2 to 13.6 KHz	Global	.5 to 2 N.M. (<1000 feet for diff. om)	Phase	Small	Moderate	• Lane Ident. • Diurnal Shifts	• Global Coverage • Low Operating Costs
Loran C	100 KHz	1000 N.M.	150 to 300 feet	Intra-Pulse Phase Lock	Small	Moderate	• Calibration • Variable Ground Conductivity • Coverage Limits	• U.S. Operated Stations • Low Operating Costs
Lorac, Raydist, Toran, etc.	1.6 to 3.0 MHz	100 to 300 N.M.	10 to 150 feet	Phase	Moderate to Large	Moderate to Heavy	• Lane Ident. Skywaves • Local Logistics	• Operating Freq. Acceptable Most Places
Loran A	1.75 to 1.95 MHz	700 N.M.	.5 to 2 N.M.	Pulse Matching	Small	Light	• Skywaves • Coverage Limits	• Inexpensive
Shoran (Long-Range)	230 to 305 MHz	200 N.M.	50 to 250 feet	Pulse	Large	Heavy	• Local Logistics • Cold Weather Can Reduce Range • Operating Freq. Not Acceptable in U.S. and Italy	
Shoran (High Freq.)	420 to 450 MHz	200 N.M.	50 to 250 feet	Pulse	Large	Heavy	• Same as Long Range Shoran • Current Lack of Broad Field Experience	• Operating Freq. Acceptable Certain Places Where Long Range Shoran is Not
X-Band Radar	9.3 to 9.5 GHz	Line of Sight to 100 N.M.	3 to 50 feet	Pulse	Moderate	Light	• Line of Sight Blockage • Range Reduced in Heavy Weather	• High Mobility

MICRO-OMEGA is a system developed by Teledyne-Hastings-Raydist as a means of achieving a maximum accuracy from OMEGA of approximately 300 feet[5].

Table 9-1 is a summary of comparative radio positioning system characteristics. It is advisable, however, to seek the advice of an expert in a geographical area where accurate positioning is required.

Navy Navigational Satellite System

"The Navy Navigation Satellite System (NNSS) provides a unique capability for highly accurate, repeatable position location in geographical and oceanographic survey."[6] This system is being used more and more in oil field operations. The NNSS consists of artificial satellites in orbit around the earth, from which world wide position fixes and latitude and longitude at predictable intervals can be obtained ten to twenty times a day. Accuracies within 110 feet have been recorded by users.[7]

The satellites are supported by tracking stations, data injection stations, the U.S. Naval Observatory, and a computing center that analyzes tracking

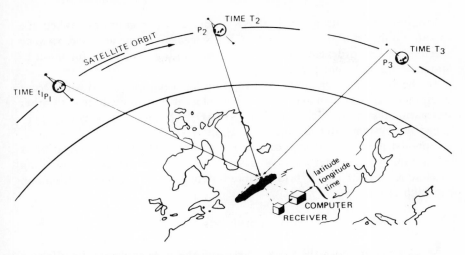

Figure 9-4. Relative geometry of shipboard user and satellite.

data and predicts satellite orbital parameters. The injection stations transmit the data to the orbiting satellite where it is stored in an electronic memory.

Each satellite transmits stable, continuous 150 and 400 MegHz signals that are phase modulated with the injected data describing the satellite's orbit. Thus, the satellite can transmit its position as a function of time. A maximum of nine two-minute data points (messages) are available during a single pass. Each two-minute message contains the predicted satellite orbital parameters as well as precise timing information. The use of this equipment only needs to:

1. Receive and process satellite data to locate the satellite.
2. Use the satellite positions thus derived, along with measured Doppler frequency shifts, to accurately locate the user's position.[5]

Reception of three or more two-minute messages during a pass permits computation of three satellite positions at each two-minute mark. Using these three positions, a foci, and the change in slant range between the user and the satellite (calculated from received data), the user's position can be determined with an accuracy of better than one-tenth of a nautical mile. Figure 9-4 illustrates the relative geometry of the shipboard user and the satellite.

There are various means of improving the accuracy of satellite navigation. Often, a real time signal processor combined with other navigating equipment will be used. LORAN-C data can be acquired and used for calibration at the option of the user.[8] The long LORAN-C data are optimally integrated with the NNSS measurements to provide highest possible per-pass NNSS calibration information for LORAN under at-sea conditions. When the offshore unit is mobile, calibrated LORAN measurements are used between satellite passes to develop a continuous best estimate of track for input to standard navigational procedures.

In the same manner, satellite navigation has been integrated with Decca equipment. It is anticipated that similar integration will be done with other navigational systems such as RAYDIST, TORAN, sonar velocity logs, gyro speed log systems and inertial navigation.[9,10]

For a drilling vessel on location, the accuracy of satellite navigation is exceptionally good after about 40 passes. Many oil companies are now using satellite navigation or a hybrid satellite system for position determination on the high seas.

Workboats

Workboats are both the work horses and the pack mules of the offshore oil industry. From their origin of war surplus LCTs (Landing Crafts-Tank) used by the industry in the Gulf of Mexico, workboats have been completely redesigned to be seaworthy, all-purpose vessels. Now, workboats are used as general supply vessels for all offshore operations, seismic exploration, seagoing tugs, and even telephone cable laying vessels. Requirements for servicing floating drilling rigs make it necessary for supply vessels to handle heavy anchors and mooring lines in addition to hauling supplies.

Thus, the modern supply boats have evolved. Chain and cable handling and storage facilities, anchor lifting gear, bulk mud and cement equipment, special tanks for potable water storage, and a large deck area are a part of these versatile vessels. Workboats used with large semis in rough weather operations are around the 200-foot, 800 ton class. Smaller and less sophisticated workboats are used in relatively calm areas where severe storms are predictable. Obviously, the geographical area as well as the load requirements will have a marked effect on workboat requirements.

The similarity between workboats and drilling vessels is that both float and both, require stability evaluations. Beyond that, the vast differences in operational requirements make an appreciable difference in their preferred characteristics (see Table 9-2).

Workboats should be able to carry *large* amounts of cargo at a *high* speed, have *good* range and be able to operate *safely*. Determining the relative need for the italicized qualities requires judgment based on operating needs.

Loading

Heavy supplies, with the exception of liquids and bulk solids, will be carried on the deck. When the deck is loaded to the maximum, the vessel

Table 9-2
Desirable Comparison of Workboat vs.
Drilling Vessel Characteristics

Property	Workboat	Drilling
Primary location	Transit	Moored
Location during storm	Maneuver or go to harbor	Moored
Vessel motion	Generally bad, but of minor concern*	Must be low
Load arrangement	Load and unload part of all of cargo, then balast to safe trim and stability	For drilling operation efficiency
Loading and unloading	Easy to load and discharge full cargo	Relatively little cargo change
Maneuverability	Very important	Relatively unimportant
Propulsion/displacement load line limit	2 to 4 hp/ton	Less than 1 hp/ton

*Special purpose vessels for North Sea use are being built with motion characteristics as a primary design criterion! Generally used with platforms and construction, these vessels are still considered specialty equipment in spite of the fact that they may be referred to as workboats.

must not be loaded to the load line. The load line limit depends on the hull strength, while the maximum deck cargo is limited by deck strength and vessel stability. For an example that presents typical data on two supply vessels, consider Table 9-3.

In Table 9-3, Ship 2 can carry 118 long tons more weight when the deck is not loaded to the maximum.

Load line data depend on the laws of physics, and thus are international. The specific load line to be used will depend on the weather areas and water density, and are coded and displayed as a Plimsoll Mark shown in Figure 9-5.

When calculating maximum deck loading, it is generally assumed that the load's center of gravity is three feet above the cargo deck. If P-tanks full of

Table 9-3
Characteristics of Offshore Supply Vessels

	Ship No. 1	Ship No. 2
Length overall (LOA)	185'	200'
Length beam (LB)	166'	185'
Depth, molded (D)	16'9"	18'6"
Lightship displacement, LT	816	844
Fuel, gal	185,000	183,000
Potable water, gal	32,969	31,250
Drill water, bbl	3,880	6,400
Length after-deck	92.6'	105'
No. dry mud tanks (P-tanks)	4 @ 1,060 ft^3	4 @ 1,125 ft^3
Gross tonnage	699	
Load line data:		
Draft (H)	14'10"	15'6"
Length water line (LWL)	166	185
Displacement, LT	1,767	2,050.4
Maximum deck cargo:		
Deck cargo, LT	394	500
Displacement, LT	1,613.1	1,932.4
Fuel oil, LT	127	395.4
Potable water, LT	62.4	56.4
Drill water, LT	207	65.55
Draft (H)	14'1"	15'
Freeboard, ft	2.67	3.50
Trial speed, kts	13.7	15

cement are placed on a boat's cargo deck, the center of gravity of the load may be higher than estimated in Table 9-3. The maximum allowable deck load of P-tanks, therefore, may be less than the value listed in the table. On the other hand, if the load is drill collars, the maximum load will probably be the listed value, because even if the center of gravity is below three feet, the deck load is limited by the strength of the deck.

Free surface (as discussed in Chapter 2) for drilling vessels can manifest itself in strange ways on a workboat. Water from waves washing over the deck can be trapped in large tubular goods, and will become a free surface. Resistance time may not be long, but captains on workboats in rough weather areas have reported that they were nearly capsized from the effect.

Large tubulars should be plugged when being transported on the open deck. Special care should be taken when transporting open tanks such as mud tanks. All tanks should be empty of liquids, and precautions should be taken to assure that they will not pick up water in transit. Cross-connected

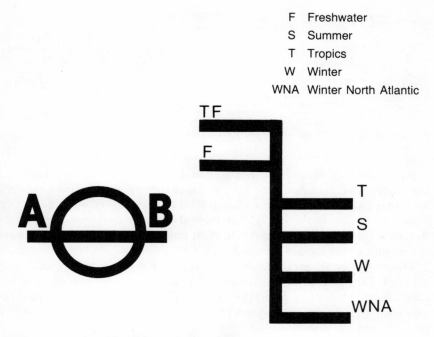

F   Freshwater
S   Summer
T   Tropics
W   Winter
WNA Winter North Atlantic

Figure 9-5. Load line markings, Plimsoll Mark.

wing tanks cause a considerable reduction in stability and are the reported reason for the loss of some workboats.

Storage capacities for supplies other than deck cargo are also important. Drilling vessels must be refueled, and adequate quantities of drilling water must be carried to the rig. The number and type of rigs to be serviced are also important when determining the most effective workboat available for your needs.

Arrangement

Arrangement refers to the location of the necessary parts of the vessel and stowage compartments. Designed arrangement has a great impact on the operating efficiency as well as on the general safety aspects of a workboat.

Workboat arrangements are the results of a series of compromises. Consequently, it is inadvisable to form fixed rules concerning what makes a *good* or a *bad* arrangement. A *bad* feature may be selected so that additional *good* features can be incorporated.

For example, it is generally cheaper to locate the engine room as far aft as possible; however, this presents problems with regulations and common

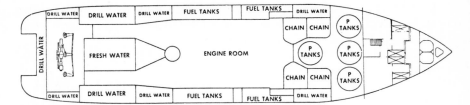

Figure 9-6. Supply boat—engine room forward, typical arrangement.

sense. Stacks on oceangoing boats should be well forward and high. An international requirement is to have access to the engine room from the deck both fore and aft; this requires a hatch behind the engine room. A hatch that is well aft decreases the boat's effective stability and could become a safety hazard if blocked open. As a result, most modern seagoing workboats have the engine room located forward.

Figure 9-6 is an arrangement of a supply boat with the engine room forward. Outboard of the engine room are fuel tanks that extend up to the cargo deck. This offers protection against flooding the engine room if the side shell is penetrated. Most workboats have double bottom tanks, which protects against flooding from grounding.

Location of the P-tanks with respect to the drill water tanks would be a disadvantage with this boat if it were to be used for more than one drilling vessel or platform. If one platform needed drill water and another the full supply of bulk material, it would be difficult to carry full loads of both supplies, offload one supply, and trim the vessel against storm conditions.

It should be possible to correct for loading variations by ballasting. This requires ballast tanks in the right places. These *right* places will depend on the vessel's use. Unfortunately, the leasee has no control over vessel design, and must make the best of what he has to work with.

Anchor Handling and Towing

Anchor handling and towing place stringent requirements on vessel size, special equipment, hull design, and stability, as well as on horsepower. Weather has the greatest affect on these requirements.

Not only does the workboat have to be sturdy enough to withstand the weather, but rough weather can also require heavier anchors and support equipment. For example, running an anchor in 3-feet seas and 10-knot winds while the vessel pays out wire rope that weighs less than 20 pounds per foot can hardly be compared to running the same anchor against 8-foot seas and 25-knot winds while the vessel pays out 3-inch chain weighing about 100 pounds per foot.

Evaluation of an anchoring and towing vessel is best left to experts in the field. However, some problems and rules of thumb may be considered when evaluating a workboat for running anchors.

In calm areas where severe storms are reasonably predictable, 2,000-horsepower vessels have been used for anchor handling in shallow water and for very light tows of short duration. A *minimum* of 4,000 horsepower is needed in rough weather areas. A-frames may be used in calm areas, but are dangerous both to drilling vessels and to personnel in rough weather. A six-foot (minimum diameter) stern roller should be used to recover anchors.

Regardless of the area, chain should be stowed in lockers and not on deck. Adequate methods for securing 100 feet or more of two-inch or larger chain on a deck have not been devised for even marginal weather.

Safety of the personnel on board depends to a great extent on the condition of the equipment being used. Anchor handling is hazardous work under the best conditions, so you should be sure that the equipment used is well maintained and adequate. A well-trained and experienced crew is equally as valuable. Crew and equipment must both be capable of doing the job.

Other Workboat Requirements

Logistics require a definition of the operation. You want to hire the most economical boat to do the job, but it must be able to do the job. Rough weather and long hauls may suggest a boat that is bigger than a boat chosen if only payload were considered. Load capacity and fuel requirements of the drilling vessel(s) will strongly influence the type and number of workboats required.

Past performance in the area is the best indication of what can be expected of a workboat and its crew. Unfortunately, this information is often difficult to acquire. Prior-leases references and the ship's logs are the usual sources of such data.

Loaded draft vs. water depth at the loading dock is occasionally overlooked in the haste to acquire a workboat. This factor must be considered to avoid unanticipated expenses. It is both embarassing and costly to hire a vessel because it is cheap and then find out that to use it will require dredging. Do not forget tidal changes when determining water depth at the docks and in the water ways.

Stability

The principles of stability were discussed in Chapter 2. The concepts of stability are the same for drilling vessels and workboats, but things generally happen faster on a workboat. Small errors can grow rapidly, to become

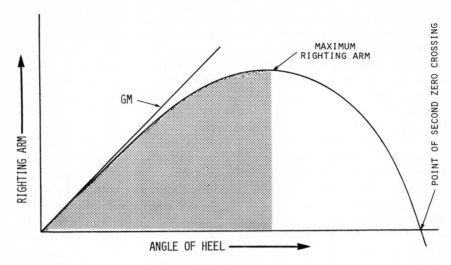

Figure 9-7. Example stability curve of a workboat with important shape factors labeled.

hazardous on a workboat in rough seas, and the situation must be corrected or the boat may be lost.

Stability curves for workboats are used to evaluate the ability of the hull to stay afloat while a problem is being diagnosed and corrected. The term *mishap stability* for a workboat might be substituted for the term *damage stability* in Chapter 2. For example, what happens to the stability when the ballast tanks are cross connected? What happens if a hatch is left open?

These questions and the many others that should be answered cannot be plotted on a single curve; however, some criteria used by major operators do aid in selecting large oceangoing vessels to resist these problems. One company's criteria for defining a stability curve shape (see Figure 9-7) is as follows:

1. Minimum GM of ½ ft
2. Minimum heel angle for maximum righting arm of 30 deg
3. Minimum heel angle for second zero crossing of the righting arm of 60 deg
4. Minimum area under the curve to the maximum righting arm of 15 ft-deg

The downflood angle can be expected to be beyond the second zero crossing for most seagoing workboats. Workboats that meet these criteria should still analyzed for individual risks by the operator.

These criteria are more stringent than are those usually required in relatively sheltered areas, where the occasion of violent storms is predictable, and vessels can be brought into harbor before being caught in a major storm. For rough weather areas and on the high seas, such criteria for the stability curve are frequently used as a first cut at workboat selection.

Inspection

A bare boat charter gives the lessee responsibility for safe and legal operation of the vessel. This means that if the boat is overloaded, the leasee will be considered responsible, just as the captain will be. Therefore, thoroughly inspect the boat before accepting it, especially if you have a bare boat charter. Use a trained inspector because he may notice a serious condition that has not been included in a check list.

It will save time to inform the captain of the extent of the inspection so that he can have the ship's documents ready and the tank tops opened and loosely in place.

The following is a guide that may be used as a general outline for inspection of a workboat.

Documents that should be in order are:

1. Load Line Assignment
2. Admeasured or Tonnage Certificate
3. Trim and Stability Booklet
4. General Arrangements Drawing and Tonnage Drawing (if the latter is available)

Before commencing the inspection, check the Load Line Assignment to be sure that the freeboards assigned are indicated on the side of the vessel by a Plimsol Mark (see Figure 9-5). The date on the Load Line Assignment should be about the same date as the *completion of vessel construction*. The date that the boat was *admeasured* should be later than the date of the General Arrangements Drawing. The dates may be close, but an Arrangements Drawing dated after the boat is admeasured is an indication that the vessel might have been substantially altered after survey. Check the first page or cover page of the Trim and Stability Booklet to be sure that a government has approved of the loadings in the book. Check to see if the maximum deck load is the advertised value. Ask the captain if the local government or the owner has issued a letter that imposes conditions or restrictions concerning the loading or operating of the vessel.

The *tanks* may be inspected next, but you will need to have with you the General Arrangements or Tonnage Drawing that was the basis of the survey of the boat. The purpose of this inspection is to determine that each tank is being used as the plans indicate.

Look for ballast tanks being used as fuel tanks. If this is legal, the plans should indicate such use. Check for ballast tanks being used as potable water tanks. Be sure to inspect the double bottom tanks for prescribed content, *unless* they are designated as fuel tanks.

If fuel tanks are not symetrically arranged, inquire as to what tanks are used for list correction, and how they are filled and emptied. Judgment is required to determine if the procedure is safe and reasonable. If tanks are being used contrary to plan, authorization for the change should be checked to be sure that the boat is operating legally.

Check assigned voids to be sure that they are not being used as tanks. The *hull* should be inspected next, with the following objectives:

1. To determine that water tight bulkheads do not have unauthorized accesses.
2. To determine that water tight doors are provided where the plan requires them and that they operate properly.
3. To look for indications that the vessel has had serious structural damage.
4. To determine the condition of fire fighting and lifesaving equipment.
5. To observe the general cleanliness of the boat. Boats with good housekeeping are generally more efficiently operated and less vulnerable to fire.

The arrangement plan is needed for the inspection and serious deviations between *what the boat is* compared with *what the plan says it is.*

Equipment operation is also important. Check the operating condition of necessary items such as the following:

1. Air compressors
2. Deck winches
3. Anchor wildcat
4. Capstans
5. Stern roller

If you are not convinced that the boat is a legal boat, ask for assistance, especially if it appears that, due to availability or significant economic considerations, you may have only one choice.

Tides

The *Encyclopedia Britannica* defines tides thus: "Sea-level oscillations of approximately daily, or diurnal, and twice daily, or semidiurnal, period are a world wide phenomenon observed at continental coasts and islands." To

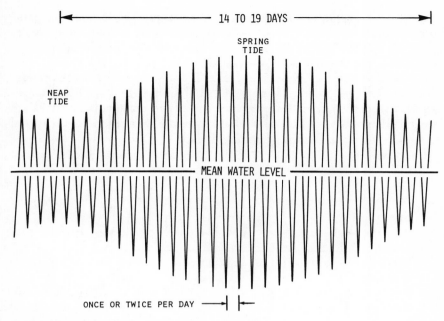

Figure 9-8. Idealized oscillations of tides.

coasts and islands, we might add vessels with risers that are attached to the sea floor.

Tidal changes are important to kelly bushing measurements and slip joint space out. All measurements should be related to a datum that is usually the *mean water level.* Tidal effects differ over the world, and all vessels used to drill on the high seas should have a set of tide tables for the area. These tables are available from the U.S. Department of Commerce or from most countries for which shipping is a major industry.

Tides are similar to an amplitude-modulated wave, as shown somewhat ideally in Figure 9-8. Tide cycles may occur once or twice a day. For example, a full tide cycle is 12 hours, 25 minutes in the North Sea, and about 24 hours in the South China Sea. Tide variations extend from about 1 foot (0.3 meters) in the South China Sea to variations greater than 50 feet (15 meters) in the Bay of Fundy.

Ranges between high and low tides will also vary at a location. The maximum range is called the *spring tide* and the minimum range is called the *neap tide.* A full cycle for the spring and neap tides vary from 14 to 19 days. Examples of tide data for some of the ports in the United States are shown in Figure 9-9.[12]

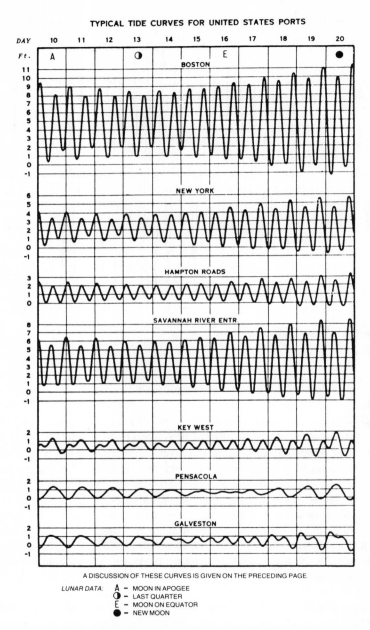

Figure 9-9. Typical tide curves for United States ports.

Where tidal changes are significant, an estimate of the tides should be determined. This requires using the data from the tide tables and modifying it to fit what is actually occuring at the well site. Data from tide tables are naturally recorded for harbors and not well sites, so some judgment is required for the correction. Some captains use the nearest harbor for the initial data. Others prefer to extrapolate data between harbors. In either case, the data will probably have to be adjusted to the well site. If you do extrapolate, remember that both time and amplitude are important, and both will require extrapolation. The tide period, however, should not change within the geographical area.

To correct the tide tables for well site use, the information should be taken frequently at first; then, spot checks should be made periodically while drilling the well. The heave indicator described in Chapter 6 can be used to take the data. Initially, six points should be taken per cycle, every two hours for a tide with a 12-hour period.

The data can be corrected based on the assumptions that the tide period is the same as recorded in the tables, but the times may be displaced, and the amplitude of the tides are proportional to the table data. Now the tides at the well site may be predicted. The time change is easily estimated geometrically by sliding the time scale in the tables by the indicated amount. The maximum and minimum tides can be changed proportionally according to the data obtained from the heave indicator. Expressed mathematically:

$$A_t = A_o \, a \sin \left(\frac{2\pi}{P} (t - t_o \pm \Delta t) \right)$$

where:

A_t = tide amplitude at any time, t, relative to the mean water level.

a = scale factor, maximum single value amplitude of tide measured on site divided by the maximum single value amplitude of the data from the tide table for the same day.

A_o = tide amplitude from the tide table taken at time t.

t = time of interest.

t_o = time at start of period, i.e., when tide is at mean water level.

Δt = time difference between on-site data and data from the table.

P = period of a cycle.

Tides will be affected by storms and high winds. Storms have a tendency to exaggerate both the high and low tides. High winds from the sea increase the high tide and the low tide; high winds from the beach tend to decrease the level of both high and low tides.

References

1. Moyer, J.R., "The Realities of Radio Positioning," OTC Paper 1788. Presented at the Offshore Technology Conference, Houston, 1973.

2. DeLerno, J., and Marchal, A.W., "Offshore Radio Positioning Systems," OTC Paper 1099. Presented at the Offshore Technology Conference, Houston, 1969.

3. Hastings, C.E., *et al*, "Modern Electronic Surveying and Navigation for Offshore Applications," OTC Paper 1395. Presented at the Offshore Technology Conference, Houston, 1971.

4. Haislip, D.T., "Loran A/Loran C, Systems Availability," OTC Paper 2362. Presented at the Offshore Technology Conference, Houston, 1975.

5. Barker, A.C., and Cottrell, G.A., "Complementary Radiolocation Techniques for Improved System Performance," OTC Paper 1675. Presented at the Offshore Technology Conference, Houston, 1972.

6. Chernof, J., "Satellite Navigation—A Valuable New Aid to Offshore Exploration," *Offshore Handbook*, vol. 2 Gulf Publishing Company, Houston, 1971.

7. Bybee, H.H., Jr., "Navigation Satellites for Geophysical Exploration," OTC Paper 1785. Presented at the Offshore Technology Conference, Houston, 1973.

8. Anon., "The ATL Model 1100A COMPUNAV/RT Real Time Navy Navigation Satellite Signal Processing System for Marine Applications Overall Description," ATL TDO 175.1.

9. Dennis, A.R., "Satellite Positioning in Navigation for Offshore Applications: Past, Present and Future," OTC Paper 2170. Presented at the Offshore Technology Conference, Houston, 1975.

10. Halamandaris, H., and Gilbert, R.K., "The Effect of Ionospheric Corrections to Positioning Accuracy Using Satellites," OTC Paper 1396. Presented at the Offshore Technology Conference, Houston, 1971.

11. "Tides," *Encyclopedia Britannica*, vol. 18, 15th edition, Helen Hemingway Benton, Chicago, 1976, Macropedia, p. 383-392.

12. *Tide Tables* (1977). U.S. Department of Commerce, NOAA, 1976.

13. Godin, Gabriel, *The Analysis of Tides*. William Clowers & Sons Ltd., London, 1972.

Drilling Vessel Check List

Inspection of a drilling vessel for possible lease or purchase requires attending to many important details. The following check list may be used as guide for the inpection of a drilling vessel. Without such a list, important items can be overlooked. The list includes data that will allow an evaluation of the important vessel components. This will facilitate estimates of the overall drilling capabilities of the vessel under various conditions.

General Information

Country _____

Field_____Block_____

Location Latitude_____Longitude_____

Environmental Conditions

    Water depth in Area: Maximum_____Minimum_____

    Maximum tide variation ± _____

    Maximum surface current_____Usual direction_____

    Air temperature: Maximum_____Minimum _____

    Water temperature: Maximum_____Minimum _____

    Estimated drilling schedule: Begin_____End_____

    Significant wave height_____Period_____Usual

      direction_____

    Maximum wind_____Usual direction_____

Logistics Information:

	Name	Distance to (Principal) Airport	Dock	Wellsite
Towns or villages	_____	_____	_____	_____
	_____	_____	_____	_____
	_____	_____	_____	_____
	_____	_____	_____	_____
Docks (include minimum depth	_____	_____	_____	_____
	_____	_____	_____	_____
Airports	_____	_____	_____	_____
	_____	_____	_____	_____
Hospitals	_____	_____	_____	_____
	_____	_____	_____	_____

Vessel Information

Basic Information:

Rig: Name _____ Contractor _____
Type _____ (semi) (ship-like) (self propelled) _____ tonnage _____
Draft Drilling _____ Draft underway _____
Maximum Deck Load _____ Basis _____
_____
Class _____ Society _____
Date of Construction _____ Flag _____
Vessel Accommodates _____ including crew of _____

Certification Papers:

	Yes	No	Date Issued
Admeasured or tonnage certificate	____	____	____ _____
Load line assignment	____	____	____ _____
Trim and stability booklet	____	____	____ _____
General arrangement drawings	____	____	____ _____
Tonnage Drawings	____	____	____ _____
Certificate of most recent survey and note and attach any exceptions	____	____	____ _____

Other certificates required by the local government:

_____ ____ ____ _____
_____ ____ ____ _____
_____ ____ ____ _____
_____ ____ ____ _____
_____ ____ ____ _____
_____ ____ ____ _____
_____ ____ ____ _____
_____ ____ ____ _____

Last drydock: Society _____ Inspector _____
    Date _____

Check ship's log for safety drills:
    Emergency: Frequency _____
    Emergency standby: Frequency _____
    Abandon ship: Abandon ship _____

Vessel Markings (legibility):

	Yes	No
Plimsoll mark: Conforms to certificate _____	_____	_____
Classing society and number _____	_____	_____
Vessel name and home port _____	_____	_____
Net tonnage _____	_____	_____

Additional requirements for vessel depending on class:

Name (Master) (bargemaster)	License No.	Country	Expires
_____	_____	_____	_____
_____	_____	_____	_____
_____	_____	_____	_____

Remarks (overall appearance and housekeeping):

_____
_____
_____
_____
_____
_____
_____

Mooring Equipment

No.	Anchors Type	Size (tons)	Type	Lines Size and Description	Condition
1	___	___	___	___	___
2	___	___	___	___	___
3	___	___	___	___	___
4	___	___	___	___	___
5	___	___	___	___	___
6	___	___	___	___	___
7	___	___	___	___	___
8	___	___	___	___	___
9	___	___	___	___	___
10	___	___	___	___	___

Extra anchors: Type _____ Size _____
    Condition _____

Chain: Wildcat type _____ Manufacturer _____
    Power _____ Chain stopper type _____
    Locker capacity _____ Length of chain on
        hand _____
    Is wildcat properly sized for chain? _____
    Condition _____
    _____
    Last Inspection _____

Wire Rope: Winch: Type _____ Manufacturer _____
    Power
Drum Size: Width _____
    Depth _____ Capacity _____

Controls: Location _____ type ___(electric) (hydraulic)___
    Visibility operation from the controls _____

Tension Information: Type _____ Manufacturer _____
    Readout location(s) _____
    Printout location(s) _____
    Maximum load _____ Last calibration _____
    Usual frequency of calibration _____

Mooring Pattern and Line Numbering System

Design Criteria for Mooring:

	Bow	Beam
Wind speed	_____	_____
Significant wave height	_____	_____
Current speed	_____	_____
Calculated forces	_____	_____
Maximum single line tension	_____	_____
with calm water tension of	_____	_____
1/3 Line breaking strength	_____	_____

Pendants:

Line: Size _____ Length _____ No. _____

Weight/ft _____

Buoys: Size _____ Condition _____

Buoyancy _____ Are they big enough? _____

Lights: Type _____ Condition _____

Remarks: _____

_____

_____

_____

Quarters, Storage and Support Equipment

Quarters:

No. of men _____ Gally _____
Recreation space _____ Hospital _____
Office space _____

Food:

meat lockers _____ freezers _____
Dry storage _____

Distilled Water:

Type _____ Capacity _____
Storage _____

Storage Capacities:

Dry bulk (sks) _____ No. of P-tanks _____ Sizes _____
Maximum distance from pods to surge tanks (ft) _____
Dry sack (sks) _____ Handling equipment _____
Capacity of hopper (sks) _____
Drilling water _____ shipping rate (bbl/min) _____

Fuel (bbl) _____
Deck storage _____
_____
Hold storage _____
_____

Deck Cranes:

1. Manufacturer _____ Load capacity _____
   Boom _____ Traverse angle _____
   Location _____
2. Manufacturer _____ Load capacity _____
   Boom _____ Traverse angle _____
3. Manufacturer _____ Load capacity _____
   Boom _____ Traverse angle _____

Diagrams of Tugger (Air Winch) Locations:

Rig Floor   Moon Pool

Main Deck (include deck cranes)

Diving Equipment:

What company recommended the equipment _____

Diving bell manufacturer _____ Identification

no. _____ Personnel capacity _____ Depth rating(ft) __

Emergancy life support time on bottom at maximum

Depth _____ Inspected by _____ Date _____

Decompression chamber manufacturer _____

Identification no. _____ Personnel capacity _____

Maximum simulated depth(ft) _____ Inspected by _____

Date _____

Individual diving gear:

	Quantity	Type	Depth Rating (ft)
1			
2			
3			
4			

Acoustic communications frequencies _____

Standby equipment _____

_____

Remarks: _____

_____

_____

Cementing Unit:

Supplier _____

Prime movers _____ Unit (Hp) _____

Rate (gal/min) _____ @ (psig) _____

Performance charts on vessel? (Yes) (No)

Heliport:

Deck area (ft^2) _____ Load capacity (lb) _____

Remarks: (local government requirements) _____

_____

Pollution Control Equipment:

    Waste treatment unit type _____
        capacity (lb/hr) _____
    Gas flare type _____ Automatic ignition? (Yes) (No)
    Oil gas burner (see test equipment)
    Chemicals and methods of dispersing _____
    _____

Subsea Television:

    Manufacturer _____ Model _____
    Frame manufacturer _____ Type _____
    Are adequate spare parts kept on board? (Yes) (No)

Other Instrumentation (Not Drilling):

    _____
    _____

Other Equipment:

    _____
    _____

Remarks:

    _____
    _____

Drilling Equipment

Depth rating of rig _____ Basis _____

Hoisting equipment _____

Derrick manufacturer _____ _____

Maximum load _____ Inspected by _____
    Date _____

Drawworks manufacturer _____ Model _____
    Prime mover power rating _____

_____

Equipment	Manufacturer	Model	Capacity	Inspection By	Date
Crown block	_____	_____	_____	___	___
Traveling block	_____	_____	_____	___	___
Drilling hook	_____	_____	_____	___	___

Drillpipe elevators:

1. _____	_____	_____	_____	___	___
2. _____	_____	_____	_____	___	___
3. _____	_____	_____	_____	___	___

Elevator bales:

1. _____	_____	_____	_____	___	___
2. _____	_____	_____	_____	___	___

Drilling Line: Size _____ Type _____

Elevator guide system condition _____

Pipe racker: Manufacturer _____ Type _____
    Capacity _____ Handle drill collars: (Yes) (No)

Rotating equipment:
    Rotary Table: Manufacturer _____ Model _____
        Size _____ Power (Hp) _____ @ (rpm) _____
    Swivel: Manufacturer _____ Model _____
        Rated load _____ Rated torque (pwr) _____

	Manufacturer	Size	Pressure Rating
Kelly 1	_____	____	_____
Kelly 2	_____	____	_____

Kelly 3 _____ _____ _____
Kelly cock _____ _____ _____
Kelly safety valve _____ _____ _____
Spare safety valves 1 _____ _____ _____
            2 _____ _____ _____
            3 _____ _____ _____
Rotary hose 1 _____ _____ _____
            2 _____ _____ _____

Power tongs: 1. Manufacturer _____ Model _____
      Size _____ Torque _____
      2. Manufacturer _____ Model _____
      Size _____ Torque _____

Casing running equipment:

	Size	Manufacturer	Model	Capacity	Inspection By	Date
Elevators						
1	_____	_____	_____	_____	_____	____
2	_____	_____	_____	_____	_____	____
3	_____	_____	_____	_____	_____	____
4	_____	_____	_____	_____	_____	____
5	_____	_____	_____	_____	_____	____
Spiders						
1	_____	_____	_____	_____	_____	____
2	_____	_____	_____	_____	_____	____
3	_____	_____	_____	_____	_____	____
4	_____	_____	_____	_____	_____	____
5	_____	_____	_____	_____	_____	____

Circulating Equipment:

Mud Pumps	Manufacturer	Model	Available Liners	Power
1				
2				
3				
4				

Circulating Pumps	Manufacturer	Model	Capacity at Zerohead	Power
1				
2				
3				

Shale Shakers:

   1 Manufacturer _____ Model _____
   2 Manufacturer _____ Model _____

Mud Tanks	Use	Dimensions	Volume
1	_____	_____	_____
2	_____	_____	_____
3	_____	_____	_____
4	_____	_____	_____
5	_____	_____	_____
6	_____	_____	_____

Ventilation for tanks below deck (change air every minute):
_____

Mud Mixing Equipment: _____
_____

Desanders: _____
_____

Desilters: _____
_____

Centrifuges: _____
_____

Degassers: _____
_____

Instrumentation: _____
_____

Diagram of Mud Pit Transfer System

Remarks: _____
_____
_____
_____

Drilling Instrumentation (include make): _____
_____
_____
_____

Ship's Power

Prime Movers	Manufacturer	Model	Control System	Shaft Power Rating
1	_____	_____	_____	_____
2	_____	_____	_____	_____
3	_____	_____	_____	_____
4	_____	_____	_____	_____
5	_____	_____	_____	_____
6	_____	_____	_____	_____

Generators	Manufacturer	Model	Rated Power Output	Output AC/DC
1	_____	_____	_____	_____
2	_____	_____	_____	_____
3	_____	_____	_____	_____
4	_____	_____	_____	_____
5	_____	_____	_____	_____
6	_____	_____	_____	_____

Voltage Inverters	Manufacturer	Model	Type	Power Rating
1	_____	_____	_____	_____
2	_____	_____	_____	_____
3	_____	_____	_____	_____
4	_____	_____	_____	_____
5	_____	_____	_____	_____

Power control system manufacturer _____

Power allocation priorities _____

Emergency generator: Power rating _____

Condition _____

Air supply: Storage capacity _____

Maximum pressure _____ Compression rate _____

at _____ pressure

Emergency air supply _____

Condition _____

Well Control Manifold

Diagram of manifold—show all valves, chokes, ells tees and targets; check pressure rating on all components.

Manifold:
    Minimum pipe size (in.) _____ Pressure rating (psi)_____
    Targets on all bends? (Yes) (No)
    Do all pipes have correct pressure rating? (Yes) (No)

Chokes:

	Choke Numbers			
	1	2	3	4
Manufacturer	___	___	___	___
Type	___	___	___	___
Pressure rating	___	___	___	___
Standpipe pressure gauge:				
Visible from choke	___	___	___	___
Accuracy	___	___	___	___
Distance to pump throttle	___	___	___	___

Remarks: _____

_____

_____

_____

Stand Pipe Manifold

Diagram of manifold: Show all valves, ells, tees and check pressure ratings on all components.

Manifold:
    Minimum pipe diameter (in.) _____ Pressure rating (psi)_____
    Do all pipes have the correct pressure rating? (Yes) (No)

Drill Pipe, Collar Inventory

Hardbanded drill pipe is unacceptable.

Drill Pipe:

Quantity	Size	Wt	Grade	Range	Tool Joint	Condition
————	————	————	————	————	————	————
————	————	————	————	————	————	————
————	————	————	————	————	————	————
————	————	————	————	————	————	————
————	————	————	————	————	————	————

Drill Collars:

Quantity	OD	ID	Range	Connection	Condition
————	————	————	————	————	————
————	————	————	————	————	————
————	————	————	————	————	————
————	————	————	————	————	————
————	————	————	————	————	————
————	————	————	————	————	————
————	————	————	————	————	————

Drilling Riser

Riser Manufacturer _____ Size: OD _____
ID _____ Riser yield (after welding and heat
treating) _____

Riser joints with field welds are unacceptable.

Standard riser joint length _____ Quantity _____
Pup joints _____
_____

Kill/Choke lines: Size OD _____ ID _____
    Pressure rating _____
Riser inspected by _____ Date _____
    Usual frequency of inspection _____
Buoyancy: Manufacturer _____ Type _____
    Buoyancy per joint _____ Basis _____
    Remarks: _____
_____

Slip joint: Manufacturer _____ Collapsed
    Length _____ Stroke Length _____
Diverter: Manufacturer _____ Can Riser be
    Shut In? (Yes) (No)
Vent lines 1. ID _____ Length _____ No. of bends _____
              2. ID _____ Length _____ No. of bends _____
Remarks on diverter system: _____
_____

Flexible joint: Manufacturer _____ Type _____
    Maximum deflection _____ Pressure balanced? (Yes) (No)

Blowout Preventer Stack

(Duplicate for two-ram stack)

Stack: Fabricator _____ Size _____ Rating _____
    Total weight _____
Rams: Manufacturer _____ Type _____
    Lock _____
Annulars: Manufacturer _____ Type _____
    Quantity _____

Drawing of Stack Arrangement:

(include kill/choke line outlets, targets overall height and distance between rams):

Kill/choke valves: Manufacturer _____ Model _____
    Quantity _____ Do any use failure mechanism instead of hydraulic pressure to close? (Yes) (No)
Kill/choke line transition around ball joint: Material _____
    Manufacturer _____ Configuration _____
Top hydraulic connector manufacturer _____
    type _____
Bottom hydraulic connector manufacturer _____
    type _____

Stack inspections:

Equipment	By	Date	Usual Frequency
Shear/blind ram	_____	_____	_____
Pipe rams 1	_____	_____	_____
2	_____	_____	_____
3	_____	_____	_____
Annulars 1	_____	_____	_____
2	_____	_____	_____
Hydraulic con.: Top	_____	_____	_____
Bottom	_____	_____	_____
Choke/kill lines, valves and Targets	_____	_____	_____

Remarks (include handling system): _____

_____

_____

_____

_____

BOP Control System

Manufacturer_____Command type? (H) (EH) (EH/MUX)
   Maximum commands available _____
Emergency backup: Manufacturer _____
   Type? (Auto) (Acoustic)
      Maximum commands available _____
      List Functions _____
   _____

   If acoustic: Subsea package responds? (Yes) (No)
      Down frequency_____Up frequency_____Battery
      life_____Subsea charge? (Yes) (No)

Accumulators:

	No.	Precharge Pressure	Per Accumulator Liquid Volume	Useable Volume	Total Useable Volume
Surface	_____	_____	_____	_____	_____
Stack	_____	_____	_____	_____	_____
Dedicated to _____	_____	_____	_____	_____	

Emergency backup:

Are surge tanks on open and close ports of annulars? (Yes) (No)
Are shuttle valves as close to function as practicable? (Yes) (No)

Reaction Times of Major Functions:

		Reaction Times (sec) (Pod)	(Pod)
Lower pipe ram:	Close	_____	_____
	Open	_____	_____
Middle pipe ram:	Close	_____	_____
	Open	_____	_____
Upper pipe ram:	Close	_____	_____
	Open	_____	_____
Stack annular:	Close	_____	_____
	Open	_____	_____
Shear ram:	Close	_____	_____
	Open	_____	_____

Hose length _____ Pod _____ Power hose ID _____
Hose length _____ Pod _____ Power hose ID _____

Remarks: _____
_____
_____
_____
_____

Motion Compensation and Tensioners

Motion Compensator:

Manufacturer _____

Model _____ Location? (Traveling block) (Crown)

Maximum load capacity _____ Pressure

rating _____ Air reservoir volume _____

Compensator displacement volume for one stroke _____

Is compensator volume less than 30% of reservoir

volume? (Yes) (No)

Calibrated: By _____ Date _____ Usual

frequency _____

Remarks: _____

_____

Riser Tensioners:

Manufacturer _____

Model _____ Quantity _____ Load capacity per

unit _____ Pressure rating _____ Angle of lines

to riser _____

Reservoir Sharing:

Tensioners	Volume Displ Both Tensioners (VT)	Reservoir Volume (VR)	Ratio (VR/VT)≤0.30
____ & ____	_____	_____	_____
____ & ____	_____	_____	_____
____ & ____	_____	_____	_____

Tension gauges calibrated by _____ Date _____

Usual frequency _____ Calibration

method _____

Tension Lines:

Size & description _____

Criteria for changing lines? (Inspection) (Ton-mileage) (Both)

Guideline Tensioners:

Manufacturer _____

Model _____Load capacity _____

Pressure rating_____Air reservoir volume _____

Displacement volume of all tensioners to one

Reservoir _____Is displacement volume less than 30% of

Reservoir volume? (Yes) (No)

Tensioners calibrated by _____ Date _____

Guideline size and description _____

Remarks on guideline maintenance and replacement: _____

_____

_____

Formation Testing Equipment

Surface test equipment:

Designer of overall system _____ Date _____
Fabricator of overall system _____ Date _____
Criteria for design:
  Oil (BOPD) _____ Gas (MSCF/D) _____
  Water (bbl/day) _____ Inlet pressure (psig) _____
  Maximum back pressure at holding tank (psig) _____
  Volume factor: Stock tank oil vol. per reservoir oil vol_____

	FABRICATED BY	PRESSURE RATING (psig)	INSPECTION TYPE	BY	DATE
Surface test tree	_____	_____	_____	_____	_____
Flow line manifold	_____	_____	_____	_____	_____
Choke manifold	_____	_____	_____	_____	_____
Heater	_____	_____	_____	_____	_____
Separator #1	_____	_____	_____	_____	_____
Separator #2	_____	_____	_____	_____	_____
Storage tank	_____	_____	_____	_____	_____
Piping	_____	_____	_____	_____	_____

Oil pump type _____ Capacity (bbl/day) @ Inlet
  pressure (psig) _____ Outlet pressure (psig) _____
Air compressor type _____ Capacity (SCF/min) @ Inlet
  pressure (psig) _____ Outlet pressure (psig) _____
Water can be furnished (bbl/day) _____ @ Outlet pressure (psig) _____
Gas burner(s): Quantity _____ Manufacturer _____
  Rated oil capacity (bbl/day) _____ @ Inlet pressure (psig) _____
  Air requirements (SCF/min) _____ @ Inlet pressure (psig) _____
  Water requirements (bbl/day) _____ @ Inlet pressure (psig) _____

Instrumentation: _____
_____
_____
_____

Remarks: _____
_____
_____

Mudline control:

Is there a tree to fit the BOP stack? (Yes) (No)
Manufacturer _____
Service representative _____
Pressure rating (psi) _____
Depth rating (ft) _____ @ (psi) _____
Remarks: _____
_____
_____

Subsurface equipment:

Local service company _____
Equipment available (list below):

Index

HARDNESS of system

W ↑   hardness ↓

T initial ↑   hard ↑

Depth ↑   hard ↓

Anchor weight ↑   nochange

MOORING PROBLEM

$\uparrow N$

INITIAL EQUILIBRIUM

$T_{LINE} = 300$ KIPS

$W_{air} = 100 \ \#/ft^3$ (CHAIN) $\Rightarrow W_{bouy} = W_{air}\left[1 - \dfrac{\rho_{fluid}}{\rho_{steel}}\right] = W_{air}\left\{1 - \dfrac{64.0}{490}\right\}$

A = chain length = 6000'

d = depth = 800'

(i) Initial angle between mooring line & horizontal

$H = T - wd \quad = 300 - 86.7(800)$

$\phi = \cos^{-1}\left(\dfrac{H}{T}\right) = 39.8°$

(ii) WORKBOAT CONNECTS @ point A & APPlies 50 K.P Northward force
· determine new location of moonpool relative to its old location
· determin resulting tension in line to north

soln: when point A moves distance d_A toward North, point B moves d_A to North (ie to B'). This results in point A moving d_A feet closer to Anchor North of A

· @ same time B moves distance $d_B = d_A \sin 30 = .5 d_A$ Further from B's Anchor

· If 50 KP force applied @ A
   - decrease in tension @ A
   - increase in lines of B & C

· ITERATE : assume 1st Approximation that horizontal component of tension @ A increases by 25 KIPS

① $H_A = 230,400 - 25000 = 205,400$ lbf
   $T_A = H_A + wd = 205400 + 69,600 = 275000$ lbf
   $V_A = \sqrt{T^2 - H^2} = (275000^2 - 205,400^2)^{1/2} = 182,855$
   $L_A = X + A - S = \dfrac{H}{w}\ln\left(\dfrac{T+V}{H}\right) + 6000 - \dfrac{V}{w} = \dfrac{205400}{87}\ln\left[\dfrac{275000 + 182855}{205,400}\right] + 6000 - \dfrac{182,855}{87}$

   $= 5791$ ft
   $\Delta L : 5801 - 5791 = 10$ ft

② If 25 KIPS taken by line north, the other 2 lines split the remaining 25 KIPS .8 12.5 Kips N-S.

$\Delta H_B \cos 60 = 12500$ lbf
$\Delta H_B = 25000$ lbf
$\therefore H_B = 230,400 + 25000 = 25400$ lbf

$T_E = 255400 + 69,600 = 325,000$
$V_B = (325000^2 - 255400^2)^{1/2} = 200987$

SUBSEA BOP STACK   TESTED @ SURFACE   w/ $P_{working}$ = 3000 psig

$V_{L-OPEN/CLOSE}$ = 250 gal

STACK TO BE INSTALLED   IN   DEPTH = 4000'

(i) Precharge psi for accumulators

- ACCUMULATOR GAS PSI of 1000 psi ABOVE PSI of surrounding reasonable assumption.
- Fluid gradient assume .45 psi/ft
- Precharge PSI = 1000 $psi_{above\ surrounding}$ + .45 × 4000' = 2800 psig

~150%

(ii) Accumulator fluid capacity required (total)

$$\text{Required fluid cap @ surface} = \frac{\text{hydraulic fluid vol}}{\text{volume factor}} = \frac{V_L * \text{API SAFETY FACTOR}}{1 - \frac{P_c}{P_{cMAX}}}$$

$$= \frac{250 * 1.5}{1 - \frac{1015}{3015}} = \frac{375}{.6633} = 565 \text{ gal}$$

@ 4000 ft

$$\text{Acc fluid capacity} = \frac{V_L * \text{SAFETY FACTOR}}{1 - \frac{P_c + P_{hydrostatic}}{P_{work} + P_{hydrostatic}}} = \frac{250 * 1.5}{1 - \frac{1015 + .45 × 4000}{3015 + .45 * 4000}}$$

∴ ITERATE	T_A	ΔL_A	T_B	ΔL_B	$\Sigma \Delta L_B$
	275	10.7	325	9.21	18.42
	270	13.06	320	7.47	14.94
	265	15.5	315	5.68	11.36
	268.4	13.83	318.4	6.9	13.81

CASING SEAL PROBLEM

IF NO LEAKS, HOW MUCH FLUID MUST BE pumped to seal to 5000 PSi

- test thru choke

  do @ NO LEAKS, FLUID = CHOKE line (compressed)

  $$V_{CHOKE} = \pi \frac{\left(\frac{CHOKE\ ID}{12}\right)^2}{4} \times Depth = \frac{\pi \left(\frac{3}{12}\right)^2}{4} \times 2000 = 98\ ft^3$$

  $$\Delta V = V_{CHOKE} \cdot C_p \Delta P$$

  $$= 98 \cdot 3 \times 10^{-6}/psi \cdot 5000\ PSi$$

  $$= 1.47\ ft^3$$

- IF seal leaks, how much fluid can be pumped befor one of casing strings likely to fail

  $$V_{ANNULUS} = \frac{\pi}{4} \left(OD^2 - ID^2\right) \times Length$$

  $$\Delta V = V_{CHOKE} \cdot C_{p_{Water}} \Delta P + V_{Annulus} \cdot C_{p_{Mud}} \Delta P$$

  use ΔP of weakest link   ie internal yield OD
  collapse for ID

  smallest value